Advance praise for

THE MYTHS OF TECHNOLOGY

"This book contributes to broadening and deepening debates on techno-logical innovation and its social significance and implications. It will be a valuable resource for students interested in a variety of subjects relating to technology policy and the management of technological change."

David Gann, Professor and Head of Innovation and Entrepreneurship,
Tanaka Business School and Civil and Environmental Engineering,
Imperial College London

"*The Myths of Technology* is a must-read for all those who want to think about technology beyond any narrow parochial perspective. It fills an im-portant gap in the literature and is to be highly recommended for both un-dergraduate and postgraduate audiences, and to policy makers and academia generally."

T. G. Whiston, Emeritus Professor, Environmental Regulation,
Roskilde University, Denmark; Honorary Professor,
Science and Technology Policy, SPRU Sussex University, United Kingdom

"This is a welcome collection on an ever-important topic—exploring the relationship between the hype and reality of new technologies. It is an ac-cessible introduction to the myths of technology that will be of particular interest to students entering this interdisciplinary field."

Flis Henwood, Professor, School of Computing, Mathematical
and Information Sciences, University of Brighton, United Kingdom

THE MYTHS OF TECHNOLOGY

Steve Jones

General Editor

Vol. 46

PETER LANG

New York • Washington, D.C./Baltimore • Bern

Frankfurt am Main • Berlin • Brussels • Vienna • Oxford

THE MYTHS OF TECHNOLOGY

Innovation and Inequality

EDITED BY
Judith Burnett, Peter Senker, Kathy Walker

PETER LANG
New York • Washington, D.C./Baltimore • Bern
Frankfurt am Main • Berlin • Brussels • Vienna • Oxford

Library of Congress Cataloging-in-Publication Data

The myths of technology: innovation and inequality /
edited by Judith Burnett, Peter Senker, Kathy Walker.
p. cm. — (Digital formations vol. 46)
Includes bibliographical references and index.
1. Technology—Social aspects. I. Burnett, Judith.
II. Senker, Peter. III. Walker, Kathy.
T14.5.M94   303.48'3—dc22   2008042692
ISBN 978-1-4331-0520-3 (hardcover)
ISBN 978-1-4331-0128-1 (paperback)
ISSN 1526-3169

Bibliographic information published by **Die Deutsche Bibliothek**.
Die Deutsche Bibliothek lists this publication in the "Deutsche
Nationalbibliografie"; detailed bibliographic data is available
on the Internet at http://dnb.ddb.de/.

The paper in this book meets the guidelines for permanence and durability
of the Committee on Production Guidelines for Book Longevity
of the Council of Library Resources.

© 2009 Peter Lang Publishing, Inc., New York
29 Broadway, 18th floor, New York, NY 10006
www.peterlang.com

Printed in the United States of America

Contents

Illustrations

Figures

Tables

Foreword

This book is about myths and technologies.

The combination seems a strange one. In their standard accounts, myths and technologies are, implicitly or explicitly, pitted against each other. Technologies are rational, material products of science, and ultimately expressions of modernity. Myths are irrational, mere stories, unscientific and essentially primitive. These standard accounts pave the way for uncritical normative uses of both terms. Technology, then, may be used as the utopian saviour of mankind – feeding the poor, healing the sick, bridging divides and bringing world peace. Alternatively, it is presented as the dystopian threat – converging food into fuel, introducing new infections, adding to the economic gap between the global north and south and providing ever more effective killing tools. Myths can be depicted as valuable expressions of human experience and symbolic understanding of phenomena that otherwise escape (scientific) explanation, or as distorted, false, unscientific or indeed 'ideological' stories about the world.

And yet, a combined analysis of myths and technologies seems so very pertinent. Many if not all myths have clear technological components, though the label 'technological' is anachronistic in many cases. In classic Greek or Hindu myths, technical contraptions abound. In the creation myths of most religions, the gods make interventions that current parlance would call 'technological'. Modern technologies allow humans to do things that in former times existed only in the realms of myth.

The starting point to investigate the relations between technologies and myths in this volume is on the side of technologies (of ICT and biotechnology especially) rather than on the side of the myths. The contributors investigate the relations between technology and society by asking questions about the myths and mythology that surround these technologies. This volume thus transcends the apparent incommensurability between the technical and the mythical domains, and it does so in a very fruitful way.

This innovative and exciting analysis of technology's role in society by analysing myths draws on recent work in the sociology and history of technology. That work has demonstrated how technology's development cannot be explained in deterministic terms, hypothesising an internal goal-oriented logic of technology, but must be interpreted as socially constructed and thus rooted in societal processes. Modern myths, this volume shows, are a gold mine for studying the values systems and the utopian and dystopian images that shape those processes. At the same time scholarship in technology studies has also conceptualised the impact that technologies have on society. This impact may extend much further and deeper than the immediate physical consequences, however important these often are: technologies also shape the hopes and frights for the future of society—and they typically do so by shaping the modern myths that govern our thinking about these futures. Myths will, to some extent, always play that performative role, but in some cases they are even actively created to exert certain political and economic effects in support of specific technologies. The critical stance of the authors in this volume helps to identify such hidden strategic dimensions of some modern myths. Such analyses form a crucial ingredient of discussions about the political dimensions of technology and innovation in society, and about justice and equality in the distribution of risks and benefits of these technologies.

This volume is very welcome – for scholarly and for political engagement with technologies, innovations and their democratic governance.

Wiebe E. Bijker
Professor of Technology and Society
Maastricht University

Acknowledgements

The research for chapter 10 'Myth and HIV Medical Technologies: Perspectives from the 'Transitions in HIV' project' was funded by the UK Economic and Social Research and Medical Research Councils, Innovative Health Technologies Programme (Grant: L218252011), as part of the project 'Transitions in HIV Management: The Role of Innovative Health Technologies'. Paul Flowers (Principal Investigator), Graham Hart (Investigator), John Imrie (Investigator), Mark Davis (Investigator), Marsha Rosengarten (Researcher) and Jamie Frankis (Researcher).

The research for Chapter 11, 'The Myth of the Biotech Revolution', was funded by the ESRC as part of its Innovative Health Technologies Programme. The authors would like to thank Surya Mahdi, Alison Kraft and Michael Hopkins. (Usual disclaimers apply.)

The editors would like to thank Mary Savigar at Peter Lang Publishing for her help, encouragement and numerous constructive suggestions in editing the book for publication.

Contributors

Judith Burnett (Ed.) is Associate Dean, School of Social Sciences, Media and Cultural Studies at the University of East London (UEL). Her research explores social change, particularly in the area of lifecourse, cohorts and generations. She has published in the area of generations and social theory, alongside education and autobiography.

Joanna Chataway is Professor of Development Policy and Practice in the Technology Faculty at the Open University. She has published widely, recently in the *Journal of International Development* ('Is it possible to create pro-poor agriculture related biotechnology?' 2005) and in *Research Policy* exploring company R&D strategies.

Erika Cudworth is Senior Lecturer in Politics and Sociology at the University of East London. Erika teaches political sociology, particularly in the areas of state theory, gender relations and radical politics in social movements. Her research interests include discursive analytics of material cultures, the political economy of food production, the gendering and naturing of space and place and human relations with non-human animals, particularly theoretical and political challenges to exclusive humanism. She is author of *Environment and Society* (Routledge, 2003) and *Developing Ecofeminist Theory: the Complexity of Difference* (Palgrave, 2005).

Mark Davis is a Lecturer in the Department of Sociology, School of Political and Social Inquiry, Faculty of Arts, Monash University. His recent research included the 'HIV transitions project'. His publications include (with P. Flowers, 2006) *Uncertainty and Technological Horizon in Qualitative Interviews about HIV Treatment,* and work exploring E-dating and HIV prevention in Sociology of Health and Illness.

Paul Flowers is Professor of Sexual Health Psychology, in the Department of Psychology at Glasgow Caledonian University. His research interests focus on understanding HIV prevention and the experiences of living with HIV. He is particularly interested in notions of sexual culture and the relational nature of sexual conduct. His recent research has also included the 'HIV transitions project' funded as part of the ESRC Innovative Health Technologies programme.

Bernard Kahane is Professor at the Institut Supérieur de Technologie et Management (I.S.T.M) in Paris where he teaches on strategy and the management of innovation and technology. He also researches on public innovation policy and strategies of high-tech organisations. His main research focus is on nanotechnology emergence.

Delia Langstone is lecturer in Innovation Studies at the University of East London. Her research interests include the social history of technological development specialising in new surveillance technologies, privacy and the social construction of the individual as a data subject.

Paul Martin is a Deputy Director and Senior Lecturer at the Institute for the Study of Genetics, Biorisk and Society (IGBIS), University of Nottingham. He has been awarded a number of large research project grants on innovation in biotechnology and the implications of the new genetics for society and the economy.

Alvaro de Miranda is Principal Lecturer in Innovation Studies at University of East London. He teaches in the areas of Political Economy of ICTs, Technology and Work and Technology and Sustainable Development. His research interests are in the relationship between technical experts and users of ICTs, the ideology of the information society and European science and technology policy.

Paul Nightingale is a Senior Research Fellow at the Science and Technology Policy Research Unit (SPRU) at the University of Sussex. His research interests include innovation in the pharmaceutical industry, technology policy, biosecurity and innovation theory.

Peter Senker (Ed.) is a Visiting Professor at University of East London. He initiated and co-edited the book *Technology and In/Equality: Questioning the Information Society* (London: Routledge 2000), to which he and some of this book's authors contributed chapters. He has published widely including (with Rodrigo Arocena) 'Technology, Inequality and Underdevelopment: The Case of Latin America' in the *Journal of Science, Technology and Human Values*, 2003.

Richard Sharpe is a Visiting Fellow at the University of East London. He has been researching, analysing, reporting on and lecturing about the IT world since 1970. His books include *The Computer World, Software Agents* and *UK IT Skills in 2003*. His articles have been published in *The Financial Times, The Times, The Daily Telegraph, The Herald Tribune* and numerous specialist publications.

Linda Stepulevage is a senior lecturer in Sociology & Innovation Studies at the University of East London. Her research explores gender-computing relations and the development of technical skills and knowledge. Recent publications include: 'Computer-Based Office Work: Stories of Gender, Design, and Use' in *IEEE Annals of the History of Computing* (2003) 25:4 and (with Miriam Mukasa, 2005) The social relations of large-scale software system implementation in *Journal of Information, Communication & Ethics in Society* 3:4.

Kathy Walker (Ed.) is Senior Lecturer in Media and Communications at the University of East London and leader of the Communication Studies degree programme. Her research interests include the innovation and development of new communication technologies, media policy and regulation, and public service broadcasting in the UK. She contributed a chapter to the book *Technology and In/Equality: Questioning the Information Society*, which addressed the implications of new technologies and the transactional television environment for audiences.

1. Introduction

JUDITH BURNETT, PETER SENKER AND KATHY WALKER

In this book, we explore some of the diverse myths which relate to technology in contemporary capitalist social formations, and how they play out in practice, particularly in the fields of information and communication technologies (ICTs), and nature, society and biotechnology. Myths are everywhere, and it seems, have never been more popular. In an age of information overload characterised by the proliferation of research and the drive to expand access to education across the world, it is ironic that human cultures retain a fierce attachment to the mystical, mythical and magical. There is nothing that the human race likes more, it seems (when it has a full belly, a few dollars in its pocket and an identity that it can call its own), than to settle back and enjoy the parallel worlds peopled with subjects and objects which are both like and unlike us and ours. From Harry Potter's invisible cloak to Sauron's all-seeing eye, old magic is put into new bottles. Enduring concepts which have structured human consciousness for thousands of years, of good and evil, man and beast and our own and other universes, are recycled, repackaged and relearned.

Technologies, and the physical, scientific and mystical world from which they are prone to spring, have a special place in myths, a long-standing appeal which is not only a matter of enjoyment but also one of awe. Awe is an integral part of mythic accounts, many of which are truly 'awful' in their horrifying features. The qualities found in potions, atoms, machines, namely their power to change the world, if not take it over, are understood in myths to be awesome. Parallel worlds are always interesting for the simple reason that they are a mirror of our own. We are uncertain in our knowledge of the world, insecure in our identities, with limited and contingent knowledge of the awesome power of the subjects and objects which people this place. The boundaries between myths and knowledge are at times slippery, and this we know, because so was it ever shown to be.

In the real world, the development and deployment of technologies have generated their own mythic structures, borrowing much older ideas and bringing these together with new ideas have produced myths repackaged for our time. Distinguishing between reality and some of the wilder claims that accompany new waves of technological development is a constant social problem, although one which is hardly new. Counter evidence has been a constant companion to our understanding of technology.

Each strand of innovation tells its own plot in the chequered history of the uses and abuses of technology. In the burst of technological innovations of industrialisation, together with beneficial instruments and surgical equipment which enhanced the quality of life and even saved some, came other innovations, such as the mass production of gun power and weapons put to use in the conquests of war and empire. The subordination of labour in the factories, railway works and grand houses of the industrial revolution was followed by gas chambers and the atom bomb whose purpose is not the preservation of life, but its elimination. The Enlightenment brought advances in human understanding which delivered reason into a world characterised in historical novella as lived by candlelight under the influence of a mixture of hysteria, hard labour and unsatisfactory drugs. But it also involved unfortunate attempts at the mastery of the universe, which carried a substantial human and environmental cost. Critical women have pointed out that many processes of subordination and death have been mainly introduced by men. Moreover, sentient people everywhere have increasingly been able to reflect upon, for example, the distinction between the characterisation of technology as a beneficial force, bringing forth a world of abundance and leisure, and the view that technologies can be deployed in damaging ways, or at least, that such simplistic characterisations deserve further examination.

For example, the 'Information Society' is consistently presented as having been created by technology, or by individual entrepreneurs immersed in innovating with technology. This concept has been widely adopted by sectors including the governmental, business, the media, and used to underpin policy approaches. In the process it has acquired all the characteristics of a technological myth, represented in an insufficiently nuanced short hand. The mythical association between technological revolutions and a better social future can be traced. For example, an important step in its establishment can in hindsight, be seen in the 1993 Delors Report for the European Commission. In this, Delors argued that the information society was the result of the dawning of a multimedia world which represents a radical change comparable with the first industrial revolution and that it can provide an answer to the new needs of European societies: communication networks within companies; widespread teleworking; widespread access to scientific and leisure databases; development of preventive health care and home medicine for the elderly (European Commission, 1993). Meanwhile, the Bangemann Report published in 1994 connected the information society with free-market ideology arguing that the creation of the information society should be entrusted to the private sector and to market forces. The not dissimilar concept of the 'information age' gained support in the United States and its history can be similarly traced. Both the 'information society' and the 'information age' character-ise technology as a positive force which can address important inequalities, and implicitly create a social revolution, in terms of lifestyle, education and so on. The policies which emerged around them emphasised the need to facilitate private initiatives and investments, and argued for wealth creation and distribution using private capital.

However, as the new technologies wave hit, other concepts also emerged, including the 'digital divide', describing a new quasi-class system consisting of those with access to, competencies to use, and knowledge of, the digital system and those lacking these resources. The two class system which would develop, of the information rich and poor, would mimic previous class formations in the sense of the wealth and opportunity gap which distin-guished them. While the digital revolution has extended the frontiers of the global village, the vast majority of the world is detached from the digital revolution and the gap between the rich and the poor among and within countries has increased. Within the United States, the country that has travelled fastest and furthest towards the 'information society', inequality has risen at the highest rates in the Western world in the course of this revolution both in terms of income and wealth. The myth of bridging the digital divide is essentially that the poor and those marginalised from the 'information society', particularly women and black people, need to be brought into it as potential customers. This, it is suggested, will simultane-ously eradicate poverty and create the conditions whereby basic needs are satisfied. According to Bill Gates, 'Because technology has the power to make such a positive difference in people's lives, we have a simple obligation: spread it' (Gates, 2000). At the same time, shrewd, aggressive private sector companies are making good profits spreading IT across the digital divide by means of providing products and services that meet the needs of the global market. Thus what we see is a complex dynamic set in train, sometimes creating, sometimes addressing and sometimes reinforcing inequalities. This doesn't necessarily shake the pole position enjoyed by myth of technology as a beneficent force, but public counter myths and more nuanced under-standings mean that the social complexities which underlie technological innovation are drawn out, and social structures and processes identified as shaping technologies, not merely the other way around.

The chapters in this book are written from a wide variety of disciplinary perspectives, and examine the boundaries between subjects and objects of technologies. We contend that many myths continue to support and sustain capitalism, generating wealth alongside the extensive and diverse forms of inequality with which it seems to be associated (Wyatt, Henwood, Miller and Senker, 2000). As part of this, myths about technology support important social tasks such as creating or maintaining social identities, communities and structures, and can be found in all of the institutional structures of society including the mass media, government and the economy. The book explores how myths and social life are brought together, and their complex interplay with social and technological understandings. The process of research and innovation is itself a contested one, the development and deployment of waves of technology themselves generate their own mythic narratives which send out important, if conflicting, messages which frame their reception and, in turn, the received view which later plays a part in their social reconstruction. The chapters explore the development and deployment of technologies, and the development and deployment of ideas about them, taking examples in specific fields to tease out the social processes at work.

Whilst this book cannot possibly be comprehensive, our chapters demonstrate the huge variety of myths of technology and how far and wide their tentacles stretch: they reach into our everyday lives on the one hand, and yet on the other stand over and above us as social ideologies, shaping our world views and governing the actions of official institutions and powerful organisations. We start in this Introduction by unpicking some of the myths about myths, following this by considering the myths of technology in broad terms.

Finally, we introduce the chapters of our book, encouraging readers to make connections in and between the different chapters, to map the tentacles of those connections and to see them as connections in the real world.

Myths about myths: The problem of being (un)conscious

Human society has always produced ideas about itself and the world which we inhabit; myths appear in all systems of thought serving civilisations and ordinary people in everyday life. Much of the twentieth-century thinking about myths has come from Levi-Strauss, Jung, Freud and Mary Douglas. The anthropologist Levi-Strauss (1966) provided some important insights into myths, suggesting that they serve to provide a self-image of society to society, an explanatory mirror for its social organisation. For example, typically the image and its explanations might provide definitions of group membership, its insiders and outsiders, its enemies and friends. It generally provides justifications and explanations for the hierarchy which is in operation, and might contain quite detailed information about the rituals which humans need to enact in order to maintain order, pacify supernatural forces and provide accounts which connect us to the past and/or to the future and so on.

One of the major insights gained by those such as Levi-Strauss was their recognition that myths work as a *system* of belief, as a *mythology*. A mythology, in contrast to a few entertaining and educational stories which satisfy us to a greater or lesser extent, is *powerful*. The social capacity to develop mythologies is, itself, an act of power, and the exercise of mythology as it is cast out and used in the world is itself powerful. While 'myths' as separate narratives may provide local and discrete explanations for why wolves eat children in the forests, or the sun rises and falls, a *mythology* may provide universal declarations for beliefs and actions[1] dictating behaviour and long-standing social organisation within specific value sets, norms and traditions. Myths in the sense used by Levi-Strauss unify the world, and present it to us as coherent and understandable in our own terms. Our world in this theory is limited by the definition of the universe on offer. This universe is the whole truth, and there is no 'outside' or alternative of which we may conceive, or about which we may wonder.

Jung (1968) enriched the debate about mythologies by developing the idea that myths are not really about the physical world of which we are conscious, but are connected to the human unconscious. Specifically, Jung argued that myths express our unconscious fantasies and desires. Myths were a long-standing feature of history. They enabled people to get by, making the world and all of our complex human relationships bearable and understandable to us. Specifically, myths help us to work out ways to behave which enable us to operate within norms expressed by the *archetypes* of human consciousness. These archetypes are kinds of projected images of the larger than life, heroic icons who serve as our role models. They are our psychosocial and psychosexual guides and are there to mop up our confused and problematical feelings. Thus, while we may feel like murdering our boss or ex-boyfriend, family member or neighbour, we are more likely to dream about it and smile unpleasantly when we meet them next. Jung argued that many of the psychic problems of contemporary society had developed because of the loss of myths and the decline in the effectiveness of these guides. This had arisen as a consequence of the development of modern social life and industrial capitalism. Modern times have brought different kinds of social relationships which are dysfunctional, creating a loss of the anchors by which people could work out who they were and how to act in relation to each other. The lack of guides which we can effectively use to work out our psychic traumas means that feelings and desires remain insufficiently processed. In other words, we are more likely to find murder and mayhem than we once were.

While Jung's theory attracts criticism in respect of some of its elements,[2] the theory that humans use symbols to express and meet their needs remains a valuable insight. His work and that of others such as Freud alert us to the possibility of myth having a role in our unconscious and conscious desires and feelings. Freud's work, famously providing an analysis of dreaming, provided ways of thinking about the role of fantasy more generally in social life. In claiming the ability to identify patterns in fantasy, irrespective of their subsequent contestation, we can see important steps being taken to show how, in our terms in this book, myths develop into mythologies, that is, systems of thought which tend to show generic elements in their various manifestations. New versions of myths may be just that – reinventions of old and prevalent ideas, which have washed around human cultures since time immemorial. It also suggests a human vulnerability or even need for myths. Mythic structures may be an integral part of our human experience. Social theories about myths have tended to show that in spite of all of our rational worldliness, education and increased cynicism, we might be more susceptible to myths that we care to think, and altogether more needy of their offerings than we care to know.

Patton and Doniger (1996) suggest that 'mythologists' have become as interested in the creators of myths and the methods available to understand myths, as the myths themselves. The search for understanding myths by Freud, Levi-Strauss, Jung and others was for master narratives which might explain the apparent universality of myth in human society, their longevity, their power. That search for a universal theory of myths was conducted within the contexts of the nineteenth- and twentieth-century social and psychological theory. But the learning curve about myths and their place in society has moved on. The emphasis today has changed, towards deconstruction of systems of myths, to relocate them in the social contexts from whence they came or flourished.

Given that myths are a consistent element of human society, a line of inquiry is around the form which myths take, in different social formations. Asking ourselves why and how particular myths develop in particular social contexts can be very fruitful. It raises the question of what other functions myths perform, beyond explaining the physical or natural world, or the rules of coexistence and survival, since it allows us to ask who makes myths and why, or why myths take the exact form that they do. This kind of social inquiry led Karl Marx, for example, to explore the powerful purposes of myths. Marx was interested in myths and saw himself as a scientist trying to work out the laws of society, rather as a natural scientist might try to work out the laws of the natural world. Marx was keen to show how ideas about society came from within society, and are deliberately produced to serve the interests of the powerful classes. He conceived myths in terms of *ideology*, a particular social form of myths which arises within specific class formations. Ideologies, he argued, are world views which not only explain the natural world and the hierarchies of the social world and lay down values and principles of action but also are deployed to further the gains of some classes or a class over others. In this version, ideologies as myths are used to deliberately hoodwink the uneducated and the poor. Ideologies are thus misleading, they preserve the *status quo* and gloss over the realities of power and social conflict. Like all myths however, they describe the known universe, making it difficult for the disadvantaged and subordinate classes to identify an alternative, or the outside. Marx's theory of revolution is intimately related to his theory of ideology, since his argument is that while it is *difficult* to become disbelieving of dominant ideologies, it is not *impossible*. Indeed, social conditions of the existence of classes in turn provide the circumstances under which alternative conceptions of society might develop, since ordinary people may come to recognise the gap between the description of social reality in the received view of the dominant ideology, and their own experience of reality, as they encounter it. The gap between the two, argued Marx, is the space in which an alternative conception of society may develop.

Today, when we approach myths, we look for the social and political processes as well as the psychic and physical. Critical theorists have built on such ideas. The Frankfurt School, for example, theorised capitalism as a society which uses ideologies in very specific ways. Herbert Marcuse (1955/1998; 1964/2002) argued that capitalism generated its own ideological systems, the purpose of which was to maintain the accumulation of capital. This was to be done most profitably not only by ensuring that workers continued to work (which had been the focus[3] of Marx's work), but that workers would continue to consume, that is buy things, including things which they didn't need. This was affected by the manipulation of desire by generating 'false needs'. This is notable, since Marx and then other thinkers present the idea that myths and mythologies can be used as tools or weapons of control by using the human capacity for unconscious thought and desire. The idea that myths may be rooted in and play back to our consciousness is one thing: the argument that this is deliberately done by powerful social forces such as governments, the media, big corporations, advertisers and so on, acting in the full knowledge of what they are doing, is another. The Frankfurt School argued that creating and using myths to invade, shape and use our fears and desires with the effectiveness which modern capitalism provides represents a new phase in the history of the use of myth in society. The twentieth century saw the emergence and perfection of a special kind of expertise in creating myths to create and manipulate consumers through markets and brands, using coded signs to convey messages which recycle the old mythic properties of immortality, comfort and fertility (see Barthes, 1972). Further work can be found in different kinds of contributions. For example, an eloquent example is to be found in Hobsbawm and Ranger (1983) who trace the process by which particular social classes and groups can invent myths about themselves, and use such 'new' traditions to codify and legitimate their social position, or their resistance to it. Ralph Samuel and Paul Thompson (1990) meanwhile in *The Myths We Live By* present an edited collection of a series of studies of the operation of myth in everyday life, in framing life stories, national identities, dominant views of gender and its roles, such as womanhood, bringing studies of myth down to a more local level, fascinating in its detail.

Nowadays, mythmakers are as likely to be found in government think tanks, major corporations both public and private, and the worlds of expert organisations (scientific, educational, medical, R&D units etc.), which provide services and knowledge (Giddens, 1991) as the traditional elites of aristocracy, sitting alongside new elites of the media and 'chatterati'.[4] At the time of writing, there is increased speculation about the power of the security forces around the world. There is considerable social anxiety about the truths and myths of what has been termed the 'war on terrorism', itself a mythic term used to describe a complex set of dynamics which has emerged over the past 10 years. The increase in the development and use of techniques and systems of surveillance, and the truth claims which accompany them, present new material from which myths are drawn, and new expert systems to feed and support social institutions called up to play in the dangerous geo-political landscapes of contemporary global power struggles.

In summary, we can say that the various ways in which we have thought about myths demonstrates their power in social systems, the extent to which they are embedded in everyday life and human experience, and their power in the institutional structures of society. We live by myths and have always done so. Yet myths are not just abstract, external entities washing around our world, to be taken up or rejected according to our personal and individual choice. This short survey has indicated that social theory, analysis and research into myths have all shown their capacity to structure and shape our world, as well as the possibility of our human agency in collusion and resistance, conscious and unconscious, willing and unwilling, as capable and incapacitated as we are by virtue of our social position. Myths are always with us it seems, and nothing in the foreseeable future suggests that this is about to change.

Myths about technology

One of the great traditions of myths is they offer characterisations and explanations of human life, and distinguish this from other categories of existence and being, animate and inanimate. An important distinction in Western thought tends to be between living or animate as opposed to those things deemed to be inanimate. This underpins the classification systems which distinguish between objects and subjects, and creates difficulties around categorising sentient but unknowing creatures, and other kinds of objects which appear to have a 'life' of their own, although not life as we know it. The Enlightenment concept of life is a specific, culturally defined one, and is one which does not accommodate different forms of life in the sense of being found in various world religions and different cultures.

For example, in Aborigine culture, the meaning of a stone which can be thought of as brought to life in the sense of an inhabitation by ancestral spirits is in contrast to the eighteenth-century system of European thought in which stones are inanimate, and part of nature, and therefore definitively cannot be said to be animate, with a Western concept of life, nor living, in any other sense. Stones became defined as belonging to nature as part of a scientific process to classify and narrate the universe. Stones became part of a classificatory universe with vast sub-structures of classification which defined and ranked all of the different kinds of 'stones' which could be identified, and their relationship to other inanimate objects. In new social theory, for example, the post-modern, this itself has become problematised. For example, stones consist of matter which changes form, and for example may incorporate the remains of living creatures in the forms of skeletons through the process of fossilisation. This does not make it less of a stone, but does make it a more complex kind of object in the world than classification tables would suggest.

The intense mapping and classification of the universe to create definitive classifications drove the development of special methodologies of investigation and classification, and travel to the furthest reaches of the European empires in a scientific mission to identify the whole of the universe (as socially constructed) at that point in time. Other systems of classification also prevail, including popular language, the media, expert systems and other kinds of systems which we might not always think of as 'systems' – cities for example, or space travel. Several new categories of objects appeared during modern industrialisation, presenting novel problems of definition. The social contexts into which they were 'born' are expressed in the form that their subsequent categorisation took. For example, stone as 'natural' is distinct from electricity which is 'man-made'. Manufactured clothing is 'mass produced', in distinction from high couture, which is 'handmade'. 'Hand-stitched' leather shoes, bags and gloves are considered desirable luxury goods, somewhat different from high street brands which carry labels that tell us their country of manufacture, perhaps 'Made in China'. As we can see, such categories are not innocent, but linked to markets, class, status, identity, economic and social inter-relationships which link the world in specific and powerful ways. Categories represent systems of meaning; they send us coded messages about norms, values, markets and power.

Thus we can see that entirely new objects, when they emerge (which from time to time they do), present difficulties to the receiving society. What sense could English society make of steam locomotives rattling through the countryside: certain death for livestock in the passing fields, and disability for passengers? Will living in the shadow of a mobile phone mast increase the likelihood of cancer, and/or reduce the value of the house with this unfortunate view? Does smoking cigarettes damage your health? As we can see, the meanings of categories are woven into narratives which could form the basis of an accurate social theory, or on the other hand, could form the basis of a fully developed myth. The difficulty lies in distinguishing between them. It was found that locomotive steam trains did *not* cause the death of cattle, but that smoking cigarettes *is* linked to the probability of the development of disease. What we learn is the unreliability of our social understanding. At what point is the distinction between myth and reality drawn, and how can we know that this is true for all time?

'Technology' became a category of interest to human society from early days and has been central to many of the stories we weave about existence. Innovations in technology present many problems of categorisation, since they are often quite novel, and have a sort of drama all of their own, partly because of the money and reputations involved, and partly because of the 'leaps' which appear to be expressed in technological waves. Technology is something tangible, the consequences of which we can see all around us, from traffic lights to digital recorders to train timetables. Unlike obscure experiments in laboratories with their outputs in learned journals read by a small proportion of the world's population, the launch of the next technological innovation 'must-have' is hard to avoid, often easily obtained, and as easily discarded when the next one comes along.

How we should conceptualise the significance of human society finding ways to reflexively develop technology and apply it, with intention, to social and economic ends has been one of the great debates in the human and social sciences, and central to much thinking about new technologies in the literature on Information and Communications Technologies (ICTs).

Humans as tool-using animals who evolved through labour and the struggle for survival represents a common narrative, which Lewis Mumford (1967) argued has attained mythic proportion in the history of ideas. The development of upright humans with big brains is generally taken to be the evolutionary break point in human history, and is marked by the time when *homo sapiens*, 'knowing or thinking humans', became fully mobile and could plan and act on intentions and needs. Mumford understands the emergence of human consciousness to mean the ability to do 'mindful' things, by which he refers to acts such as dreaming, having ideas, imagining things, developing and communicating thoughts, constructing narratives such as myths, and yearning for and express aesthetic awareness. Contrary to many classical assumptions that this had developed through learning to labour or work in order to ensure survival, Mumford argued that it hadn't just been the result of a 'long apprenticeship in making tools and weapons' (ibid.: 23), but was the outcome of the multiple experiences which humans developed the capacity to have, and which they became increasingly capable of seeking out, learning how to do, and ultimately choosing to do.

An example Mumford took was that of the hand. In classical theory, the hand is commonly thought to have developed its capacities to bend and grip principally as a tool of labour, for example to pull food towards our mouth or to manipulate implements for killing and preparation. Yet Mumford argues that the hand was used for many other functions, it 'stroked a lover's body, held a baby close to the breast, made significant gestures, or expressed in shared ritual and ordered dance some other more inexpressible sentiment about life or death, a remembered past, or an anxious future' (ibid.: 7). The hand did many things apart from just work, and enabled us to develop emotional and symbolic meanings to give to experiences.

The myth of human evolution as being the result of making and using tools has resulted in our thinking of humans as primarily 'tool-using' animals, but in fact we may be better advised to see humans as primarily 'mind-using' animals. Indeed, our minds are so powerful that Mumford suggested that ancient people faced a problem, since 'his [*sic*] mind was not entirely on his work... [since he was unable to prevent] ... the queer things he found passing through his mind' (1967: 48). The advent of humans with large brains created spare processing power, since the everyday problems of survival absorbed but just one part of all the creativity and thought available. With all this thinking power on their hands, dreaming, thinking, feeling, wondering, theorising, imagining, explaining, problem solving, designing and inventing became intrinsic parts of everyday life for the human being. Thus, for Mumford, the creation of myths was part of a wider evolutionary process, and can itself be theorised as a means for bringing dreams under control, providing explanations of world and the things which happen to us. The greatest myth of all for Mumford, however, was the myth of the machine making human as the creator of the neutral toolset. Mumford identified the toolset as being far from neutral, but applied with intentions. Mumford's analysis assumes planning and its implementation are both possible. But contemporary thinking would suggest that in a complex fluid world, planning and its implementation are processes which are liable to substantial impact from unexpected forces, and that unintended consequences are as likely to occur.

As a preliminary step in a strategy to begin thinking about the myths of technology, we might group the myths around two poles in the narratives. Our two simplified narratives are

1. technology is the answer to all of our social, economic, political, medical ills etc.,
2. technology will bring death to millions and is the harbinger of the destruction of civilisation.

Table 1.1. Myths of technology

Social fact	Meaning (1)	Meaning (2)
The mass production of PCs and the increase in availability of software which can be easily learned	New technology is cheap and easily accessible, it challenges power inequalities and enables greater participation to more people	New technology reinforces social divisions and inequalities since not everyone can access it
The new technology has/will transform the workplace	The technology will and is replacing the old division of labour which produced an oppressive class system, breaking down the hierarchies of work, with flatter management structures, the increased use of networks, more transparency in information and decision-making etc.	The hardware products are mass produced in sweat shops in the developing world by people earning extremely low wages, using designs which have often been produced in developed areas of the world by high earners.

This generates different classes and kinds of workers. The resulting labour market is unevenly distributed, with the emergence of both local and global elites, which cannot be said to be in any sense a simple distribution between developed and developing regions. |
| Technology does not rely on coal and steel etc. | This clean technology is a step forwards on the resource hungry, polluting technology of the industrial revolution. People can work in clean environments and don't die from nasty industrial diseases. | There is no easy means of disposing of PCs etc., and over-production, constant change and fashion etc, will drive us to produce PC mountains, filling excessively large landfill sites.

PCs run on electricity and nuclear power. |

These we might counterpoise as utopian and dystopian in orientation and set out in Table 1.1.

Thus, as with all myths, myths of technology offer us a view of the world in which we live, both explaining why things are as they are, legitimating the status quo, and providing us with implied codes of behaviour, if we are to survive this brave new world. In terms of social theory, what we might say is that it provides us with an ideological view of the world, it is a *pseudo-social theory*. Such pseudo and implied theory abounds in the everyday talk about the new world, with its new technology, network society, this revolution that we endure, as powerful as the first industrial revolution, and possibly as important as the first human revolution, when humans stood upright and became sentient beings. This revolution of technology has produced new social imaginaries, and carries with it powerful messages which tell us what our society is like, and explains and justifies the hierarchy which underpins it, see Table 1.2.

Table 1.2 provides a summary of the specific myths which have arisen in the context of technological revolution. As Mosco (2005) argues, the recent wave of innovation has unleashed a new wave of belief which promises our redemption, as we, the human, may leave the confines of our body and the Earthly world, for the virtual world of mutable identities told by texts.

Table 1.2. The myths of technology: What is our technologised society like?

- 24/7 – our society never sleeps.
- Connected and global – there is no one and nowhere 'outside'.
- We are moving towards a situation of near equality between the sexes.
- Medical technology can/has lead to cures and the prevention of disease.
- Our lifespan has been extended, and will continually be so. We may even live forever.
- There is constant social change.
- Time and space have changed, and we can talk to anyone anywhere at anytime in the twinkling of an eye, that is, we can transcend our bodies.
- Technology, nature and society are all different and separate.
- We are at the mercy of the machine.
- Hierarchies have become flattened, and we live in a networked social world with distributed power and decision-making.
- The new age is more democratic, open and accessible to all, with more opportunities to exercise choice and participate in decision-making.
- We can control deviant and destructive behaviour (such as criminal, and its particular categories, e.g. terrorist) by technology.
- We can address the division of the world between developed and wealthy and the developing world by taking the technology of the developed world and diffusing it to everyone, everywhere.
- We can control nature (such as the production of crops) by technology and at least manage if not defeat the seasonal cycle of growth, fertility, decline and death.

We question this, which is not to suggest that new technologies have brought only misfortune, but rather to critically engage with the fascinating social, economic and political processes which are shown to be at work.

Mosco (2005) tracks through the booms and busts of new technology innovation and investment, showing how they have driven intense economic activity and speculation. This, he argues, is not so much the work of the rational man characterised in classical economics, but rather of irrational men and women, whose belief in the digital sublime is such that the lives, reputations and vast fortunes of global corporations as well as singular actors have been staked. The myth is that the roulette of venture capitalism will provide enough winners elevated by the magical properties of digital and genetic code to the media lists of the super-rich to make it worth the risk. So was it ever – as is the 'bust' which history shows so often follows in its wake.

Approaches to technology

In terms of conceptualising ICTs in particular, much of the literature can be said to be written from a stance of which Heap, Thomas, Einon, Manson, and Mackay (1995) suggest there are three main positions:

- Version 1 – Technology is neutral; it is up to us how we use it.
- Version 2 – Technology is produced by powerful agents and it then determines and shapes our human and social relationships.
- Version 3 – Technology reflects an interplay between structure and agent, culture, values etc.

As we can see, how theorists approach thinking about the myths of technology is itself shaped by a range of different positions, which see the world differently. This raises the issue of the extent to which scientific exploration such as Marx aspired to, and our own explorations in seeking understanding today from the range of positions which we take, can also be set within the context of producing myths about technologies. While each of the authors in this book seeks understanding and critical awareness, nonetheless we are left with something of a problem. Segal in Patton and Doniger (1996) argues that scholars of myths themselves 'cannot avoid becoming mythmakers, deeply implicated in the narrative project, and not by any means the outsiders that they may present themselves to be' (6).

In spite of ourselves, or possibly because of ourselves, myths are positioned by the process and structure of Enlightenment, the age of reason is assumed to be subordinate to science. This is partly because modern science 'does so well what myth had long been assumed to do: explain the origin and operation of the physical world' (Patton and Doniger, 1996: 82).

We might need to consider that this is itself something of a mythologising of science. The 'sociology of knowledge' debate has itself debunked science as the arbiter of truth *per se*, and has raised issues of the *social* and *material* character of science, for example, in the production of science (Bruno Latour and actor networks), of knowledge paradigms (Thomas Kuhn). Yet, social scientists share a similar fate, attempting to draw understandings into the social operation of the world; they suffer from accusations of creative writing on the one hand, and on the other, of falsely separating the object of interest, 'society' from everything else although in real life they cannot be separated.

Of the status of the chapters in this book, we leave it to our readers to decide. Suffice to say, that it is our intention to analyse the myths of technology, rather than to add to them.

The structure of the book

Part I Myths about technology and inequality

The next three chapters consider relationships between myths and technology. In order for capitalism to be sustained and to continue its dominance of world economies, it needs constantly to promote myths which legitimise it – which have the effect of making sure that most people believe that capitalism and the way in which it develops and applies technology are the best conceivable form of economic organisation.

Claims are often made that new technologies offer solutions to poverty and starvation, can improve health, food, learning. Global satellite networks are proposed as means of increasing the access to information and knowledge of people living in developing countries by the use of the Internet and mobile phones and acting as a powerful stimulus to economic growth. Genetically modified organisms (GMOs) will help feed people in countries plagued with food shortages or vitamin deficiencies. Gene therapy or stem cells will be used to cure an unlimited number of diseases. Nanotechnology will create the possibility of developing materials which will lead to the creation of alternative energy sources.

de Miranda points out that technological myths are often used as ideological weapons. Technological determinism is an essential component of such myths. Technology is represented as an autonomous and rational agent of social change impervious to human control, and this masks the existence of political interests behind the changes. The myth of bridging the digital divide is used to try to allay public anxieties over increases in inequality.

For products and services to be exploited profitably, it is a prerequisite for individuals or organisations to be able to justify their claims to own those products and services. For several centuries, knowledge and successive technological developments in the forms in which they have been encapsulated, presented and disseminated have been converted into property to be exploited for profit. But now there is a vast volume of knowledge in the public domain. Search engines appear to search the sum total of human knowledge available in a digital form. Sharpe however, claims that it is a myth that men and women alike, rich and poor alike, regardless of where they are located, now have relatively equal access to the tools of the networked society.

Kahane points out, however, that the ability of technologies to achieve their suggested potential is often challenged by opponents using myths. He identifies five mythical figures linked to Western civilisation with roots in Hellenistic and Judaeo-Christian culture which are associated with technology-related risks He suggests that technological waves are a characteristic of our era; that we may benefit from understanding the mythical stories mobilised in the context of previous new technology waves; and that it is likely that the same set of mythical stories will relate to future new technology waves and to their implications for society.

Part II Myths about information and communication technologies

The book then proceeds from these general analyses of the role of myth in society and the economy – and specifically in capitalist society – to the role of myth in creating and sustaining investment in specific new technologies and, at the same time, legitimating the directions in which capitalism exploits those technologies. Chapters by Sharpe, Stepulevage, Walker and Langstone consider myths related to Information and Communications Technologies, and these chapters are followed by chapters which consider medical and agricultural technologies including biotechnologies.

Sharpe suggests that central to the myth of the continued beneficial impact of ICTs is the notion of one more push: the problems of yesterday will be solved by the adoption of today's technology. For example, it is only a matter of getting the price down far enough for every family in the consumer market to be able to afford a PC and thus to have the educational advantages which used to be available only to a few with access to computing and storage power. But the PC is no longer enough. Access to networks has to be added as the content shifts from delivery through storage media such as tapes and floppies to delivery over networks. Then dial-up network access became inadequate and broadband access became essential. But Sharpe argues that there is no stable state which will level out inequalities because replacement technologies always create alternatives.

The number of students studying at Higher Education Institutions (HEIs) in the United Kingdom has increased, and there are declining *per capita* resources. The HEIs' response to these pressures has involved moving away from collegiality, characterised by consensus building and a democratic decision-making process within committee structures; and towards direction from the centre, characterised by an information strategy based on a mechanistic rational approach derived from the private sector. Stepulevage suggests that this push for formalisation can be seen as setting the framework for ERP (Enterprise Resource Planning) with its promises of cross-functional information integration and free internal information flows. These features are highly desired by HEI managers. The study reported in this chapter focuses on myths that ERP can become a central actor in work systems by following some set of rationally determined procedures.

Mass communication through broadcasting has been mainly a one to mass medium of communication, lacking in any significant level of interactivity. But the advent of new technologies, which facilitate greater levels of interactivity between the producers of television content and their audiences, has given rise to widespread discussion, and, in some instances wild speculation about the potentials of interactive television to liberate audiences from the constraints of passive viewing. Walker concludes that rather than empowerment, an illusion of participation through interactivity is being utilised to underpin an increasingly voyeuristic mode of television which serves the increasingly competitive demands of the television market and the need to target effectively an elusive young television audience.

Langstone presents a case study of the use and promotion of closed circuit television (CCTV) in a London borough. CCTV is now widespread throughout the United Kingdom, and it often receives glowing endorsements in the media with reductions in crime figure attributed directly to its use. Newspapers often run headlines which emphasise the role of CCTV in leading to arrests with little mention of the other essential interventions. CCTV cameras do indeed detect crime by their ability to provide a record of people actually committing crimes, and some people may be deterred from committing crimes simply because they are aware that they are 'being watched'. But there is evidence that CCTV may be used to unfairly target certain sections of society such as ethnic minorities, youths and the homeless.

Part III Myths about nature, society and technology

Some theorists have suggested that technological intervention in 'nature' reflects relations of human social domination and inequality based on relations of class, gender, 'race' and postcolonialism, and is thus problematic. Extensive human manipulation of 'natural' processes may be problematic *per se*. Others have argued that human societies have always transformed nature by their technologies. Cudworth argues that our embodiment is an important and persistent form of distinction from machines.

HIV is a global epidemic. There have been significant achievements since Highly Active Anti-Retroviral Treatment (HAART) was introduced in the mid-1990s in advanced industrial countries such as the United Kingdom. Researchers found that by addressing several aspects of viral replication at once, viral activity could be slowed so that the immune systems of people with HIV could recuperate. By 1997, the combination approach to treatment had been heralded as something approaching cure. But Davis and Flowers suggest that living with HAART is challenging and difficult. This chapter explores mythology in everyday life, and in considering the hopes and fears awakened by myths, draws our attention to the psychosocial dimension of myths as part of our lived experience.

The existence of a medicinal 'biotech revolution' has been widely accepted and promoted by academics, consultants, industry and government. This has generated expectations about significant improvements in the drug discovery process, health-care and economic development that influence a considerable amount of policy-making. But according to Nightingale and Martin rather than producing revolutionary changes, medicinal biotechnology is following a well-established pattern of slow and incremental technology diffusion.

There is some evidence that political solutions to agricultural and social problems may be more successful in helping the poor than adopting technological fixes such as the Green Revolution or biotechnology. Senker and Chataway suggest that it is a myth that agricultural development, and in particular genetic modification, is driven by the motivation of feeding the hungry. Technological change directed at the improvement of the standard of living of the many millions of subsistence farmers could have very substantial effects in terms of feeding the world's hungry. But this has very low priority for major corporations, as potential markets are perceived as too small and heterogeneous to justify significant expenditures. And developing countries lack sufficient indigenous scientific and technological capability to pursue such goals on their own. There may, however, be potential for some Public Private Partnerships to achieve significant successes in agriculture – as they have had in health where they have been helpful in the development for vaccines for neglected diseases.

In conclusion, we suggest that capitalist society relies on mythology for its very existence, while those who oppose technology also rely on powerful myths to support their arguments. As each new major technology cluster is envisaged, developed and applied, similar patterns seem to emerge. Entrepreneurs use myths to create high expectations to get access to the resources they need to develop and use technology. These myths are based on technological determinism which envisages technology as an autonomous force which drives social and economic change in directions which humankind is powerless to affect, and often embody the promise that new technology will reduce inequality.

Notes

1. Such as human sacrifice.
2. For example the rather elitist view that only a small minority of the human race may be capable of transcending the need for traditional myths.
3. Although not the exclusive interest, since Marx recognised that consumption drove demand in capitalist systems.
4. Chatteratti – the chattering classes.

Part I   Myths about Technology and Inequality

The first part of this book explores the myths of technology in terms of their claims to bring inequalities to an end or, at least, to provide significant progress in this direction.

The first chapter by de Miranda explores some of the key theoretical approaches to the study of technology and innovation, in particular focusing on technological determinism. de Miranda examines the concepts of the 'information society' and the 'Digital Divide', both of which are in widespread use. He traces the history of these concepts, where they came from and how they have been used, linking these to the specific qualities of the capitalist economic formation.

In the second chapter by Sharpe, the knowledge-based economy is examined to find that the myth of the end of inequality is indeed mythic, or aspirational, rather than a reality. However, the discussion draws out that inequalities are formed by complex processes which define property and intellectual rights, some of which stem from the modern period of industrialisation, but some of which can be traced back to earlier times. This chapter considers the role of patents, rights and property in the development of the knowledge economy, showing how the legacy of earlier concepts of property has shaped the concentration of wealth that has been generated by the new technology wave.

The third chapter by Kahane steps back to consider the nature of research and development as a social process. Kahane comments upon how innovation seems to occur in waves, and how myths of technology are part and parcel of the social organisation of such waves, making this an enduring feature of contemporary technological change.

2. Technological Determinism and Ideology: Questioning the 'Information Society' and the 'Digital Divide'

ALVARO DE MIRANDA

Preamble

The research for this chapter was prompted by an event in 2002 involving a recent graduate of the New Technology degree of the University of East London's Department of Innovation Studies. She had obtained temporary employment with a London borough on a project funded by the European Union. Her work involved creating an IT room in a run-down housing estate used largely to house refugees. The estate was rat infested and the flats had water running down the walls. There was money to install the latest computers but no money to get rid of the rats or of the humidity in the flats. She wondered if this made any sense.

Introduction

The 27 September 2004 issue of *Business Week* featured a cover story titled 'Tech's Future'. Both the cover and the story were illustrated with of pictures of dark-skinned women. The one on the cover was of an inhabitant of Recife in the poor North East of Brazil described as a 'prospective PC buyer'. The main story was illustrated by a full-page photograph depicting an Indian woman, Neelamma, a 26-year-old village woman from Andhra Pradesh, dressed in a traditional sari decorated with a garland of flowers holding a Hewlett Packard digital camera. The message in the story was driven home by a large font subtitle stating, 'With affluent markets maturing, tech's next 1 billion customers will be Chinese, Indian, Brazilian, Thai'. This message was illustrated by the case story of Neelamma, who, as part of an experiment organised by Hewlett Packard, was charging local villagers '70 cents apiece for photos of newborns, weddings and other proud moments of village life' taken with a digital camera and printed with a portable printer powered by solar charged batteries which had been rented from Hewlett Packard for $9 a month.

The same theme had already been broached by *Business Week* earlier in the year, in the issue of 28 June, in a featured article eulogising India's 'digital revolution'. Its title, 'The Digital Village', contained implicit allusions to both Marshall McLuhan and Nicholas Negroponte. The main message was expressed in a quote from C.K. Prahalad, described as a leading management theorist who studies developing markets: 'If you can conceptualize the world's 4 billion poor as a market, rather than as a burden, they must be considered the biggest source of growth left in the world.'

Both in the pictorial metaphors and in the textual message, *Business Week* was presenting a particular strategy for bridging what has become known as the 'digital divide'. The poor and those marginalised from the 'information society', particularly women and black people, need to be brought into it as potential customers rather than as human beings with needs. This, it is suggested, will simultaneously eradicate poverty and create the conditions whereby basic needs are satisfied.

The strategy promoted by *Business Week* had been advocated in December 2003 at the 'World Summit on the Information Society' (WSIS) in Geneva organised by the United Nations at the behest of the International Telecommunications Union. To justify the summit, the WSIS web site cited the existence of the

> digital revolution...fired by the engines of Information and Communications Technologies has fundamentally changed the way that people think, behave, communicate, work and earn their livelihood...forged new ways to create knowledge, educate people and disseminate information...restructured the way the world conducts economic and business practices, runs governments and engages politically...provided for the speedy delivery of humanitarian aid and healthcare, and a new vision for environmental protection. [Further], access to information...has the capacity to improve living standards for millions of people around the world [and] better communication between peoples helps resolve conflicts and attain world peace.

But the site also points to the paradox that while the 'digital revolution has extended the frontiers of the global village, the vast majority of the world remains unhooked from this unfolding phenomenon' and 'the development gap between the rich and the poor among and within countries has also increased'. The purpose of the World Summit was therefore to discuss ways to bridge the digital divide and 'place the Millennium Development Goals on the ICT-accelerated speedway to achievement'.

In its 'Declaration of Principles' the WSIS declared that its purpose was

> to harness the potential of information and communication technology to promote the development goals of the Millennium Declaration, namely the eradication of extreme poverty and hunger; achievement of universal primary education; promotion of gender equality and empowerment of women; reduction of child mortality; improvement of maternal health; to combat HIV/AIDS, malaria and other diseases; ensuring environmental sustainability; and development of global partnerships for development for the attainment of a more peaceful, just and prosperous world.

The Chair of the conference, Swiss President Pascal Couchepin, claimed in his closing address that the summit had created a new political concept of 'digital solidarity'.

However, the general tone for the summit – at which the business community was strongly represented – was set by the UN Secretary-General in his keynote speech:

> The future of the IT industry lies not so much in the developed world, where markets are saturated, as in reaching the billions of people in the developing world who remain untouched by the information revolution. E-health, e-school and other applications can offer the new dynamic of growth for which the industry has been looking.

This example is typical of the portrayal of the relationship between technology and social change by the media and by policymakers. There are a number of different aspects to this representation. The first is that it ignores the fundamental nature of the creation of technology by human society. Instead it reifies technology, which in this representation acquires a 'phantom objectivity' as an agent of social change, 'an autonomy that seems so strictly rational and all-embracing as to conceal every trace of its fundamental nature: the relation between people' (Lukacs, 1971: 83).

The second and inter-related representation involves the myth of the 'technical fix', the implicit assumption that technology provides the only feasible solution to complex social problems. Thus the World Summit on the Information Society, in its Declaration of Principles, implies that information and communication technologies possess quasi-magical powers to provide solutions to the world's greatest social and economic problems such as poverty, disease, illiteracy, race and gender discrimination and environmental pollution.

The third is the use of myths about technology in order to promote particular policies and help create particular ideologies. The reification of technology, by creating the impression that the technological change is a rational, objective and inevitable process which is driving social change, hides the social forces and social interests behind the change and the fact that there are winners and losers in the process. In this case the new technologies are associated with the neo-liberal 'free-market' ideology and the combination is presented as creating a process in which everybody wins: the transnational corporations find new markets and the poor find new ways to improve their conditions – by making money from other poor people.

The next section explores the way the concepts of the 'information society' and of the 'digital divide' became influential in shaping policy and demonstrates their technologically deterministic content and ideological role. The final section examines the technological determinist nature and mythical character of the concepts, and analyses the role that they play in mystifying social reality and draws some conclusions.

Creating the myth

The information society

One of the most recent examples of an implicitly technologically determinist presentation of the relationship between social and technological change can be found in the genesis and development of the concept of the 'information society'[1] and in how it has been used by policymakers, business leaders and the media.

The concept originates in academic attempts to understand and explain the changes that took place in the economic structure of the most advanced capitalist economies in the late twentieth century. In this period, the traditionally most dynamic economic sector of the industrialised countries, manufacturing industry, went into decline, accounting for a progressively smaller proportion of both employment and of the GDP of those countries. This was accompanied by a rise in the importance of the service sector which now accounts for more than 70 per cent of employment. Study of the nature and causes of this phenomenon led to the development of a number of inter-related concepts such as the 'post-industrial society', the 'service economy' and, most recently, the 'information society' and the 'knowledge-based economy'. The exact meaning and usefulness of these concepts has been the subject of extensive academic debate.

However, the concept of the information society has moved beyond the realm of theory and has attained the status of an incontrovertible reality and a normative character. It has been widely adopted by politicians, with the support of business and the media, as the basis of policy. In the process it has acquired all the characteristics of a technological myth. The information society is consistently presented as having been created by technology, the product of the convergence of new digital information and communication technologies. The myth, as I shall demonstrate, is nearly always portrayed in utopian terms. The development of the information society is equated with progress towards a better society in which social problems will be solved by technological means and in which human beings will be both better off and more free. This alleged freedom is ascribed to the power of the technology, the de-regulation of markets and the rolling back of the influence of the state – cornerstones of neo-liberal ideology. Often liberalising and opening up markets is presented as being necessitated by technological change.

The 'information society' as a normative concept is essentially a European construct: it was originally adopted as a central plank of the influential 1993 Delors Report written by a committee led by the then European Commission Secretary-General, Jacques Delors, a French Christian Socialist (European Commission, 1993). The report argued that the development of the market economy has a decentralising effect and 'set(s) free the dynamism and creativity inherent in competition' (ibid.: 10). It further maintained that the combined effect of this and of the use of the new technologies was 'taking Europe towards a veritable information society' (ibid.: 11). The mythical association between technological revolutions and a better social future was established by the report's contention that the information society was the result of the 'dawning of a multimedia world (sound–text–image)[2] (which) represents a radical change comparable with the first industrial revolution' and that it 'can provide an answer to the new needs of European societies: communication networks within companies; widespread teleworking; widespread access to scientific and leisure databases; development of preventive health-care and home medicine for the elderly' (ibid.: 10).

Soon afterwards, the concept was consolidated into European policy by the establishment of a High Level Group of Experts on the Global Information Society, composed almost exclusively of senior executives of Europe's leading computing and telecommunications firms. Martin Bangemann, European Commissioner responsible for industry and telecommunications, was appointed to lead the Expert Group and their report, which became well known as the Bangemann Report, was published in 1994 (European Council, 1994). The report opens with a clear expression of the technological myth and its utopian promise:

> Throughout the world, information and communications technologies are generating a new industrial revolution already as significant and far-reaching as those of the past.
>
> It is a revolution based on information, itself the expression of human knowledge. Technological progress now enables us to process, store, retrieve and communicate information in whatever form it may take – oral, written or visual – unconstrained by distance, time and volume.
>
> This revolution adds huge new capacities to human intelligence and constitutes a resource which changes the way we work together and the way we live together. (European Council, 1994: 3)

The association of the information society with free-market ideology is also made explicit in the report. The European Commission press release announcing its publication highlighted this fact:

> The creation of the information society should be entrusted to the private sector and to market forces. Existing public funding should be directed to target its requirements. At Union level, this may require reorientation of current allocations under such headings as the Fourth Framework Programme for research and development and the Structural Funds. (From: News Release on the Bangemann Group report on the Global Information society: European Commission Spokesperson Service)

The Bangemann Report was adopted by the European Union Heads of Government at the European Council meeting in Corfu in June 1994.

The concept of the information society did not have much influence in the United States. However, a parallel dominant normative concept was developed there simultaneously, that of the information age. This derived largely from the influence of Alvin Toffler's 1980 book *The Third Wave* over leading US politicians of both main parties, notably Al Gore and Newt Gingrich. Newt Gingrich, Republican Speaker of the House, had been a friend of Toffler since his days as a junior academic at West Georgia State College. In 1975, at the request of Congressional Democrats, Toffler organised a conference on futurism and 'anticipatory democracy' to which Gingrich was invited. Toffler later described in his next book *Creating a New Civilization: The Politics of the Third Wave* how the conference led to the creation of the Congressional Clearing House on the Future, a group which eventually came to be co-chaired by a young senator named Al Gore (Farrel, 2001). In *The Third Wave*, Toffler had predicted that the future would be shaped by the new information technologies. In the introduction to his previous book, *Future Shock*, Toffler had made clear that the purpose of predicting the future was to help citizens adapt to it. Toffler's work thus established the mythical proposition that the future was technologically determined and therefore inevitable. Citizens will have no option but to adapt to it, lacking the power to exert any influence or control. The message was taken up enthusiastically by both Gore and Gingrich.

Fired by the wish to facilitate the development of the information age, Gore made the creation of the United States's National Information Infrastructure (NII) and of a Global Information Infrastructure (GII) his first priority after he became Clinton's Vice-President in 1992. The development of policy on the information infrastructure was entrusted to an Information Infrastructure Task Force (IITF) operating under the aegis of the White House's Office of Science and Technology and of the National Economic Council and on which all the key government, regulatory and industry players were represented. The IITF developed an Agenda for Action on the

National Information Infrastructure. According to this, the development of the NII could 'help unleash an information revolution that will change forever the way people live, work and interact with each other'. People would be able to 'live almost everywhere they wanted, without foregoing opportunities for useful and fulfilling employment, by "telecommuting" to their offices'; 'the best schools, teachers, and courses would be available to all students, without regard to geography, distance, resources, or disability'; also health-care systems 'could be available on-line, without waiting in line, when and where you needed them' (IITF, 1993).

The US government took the proposal for the creation of a Global Information Infrastructure to the world stage. In March 1994, Al Gore addressed the conference of the International Telecommunications Union (ITU) in Buenos Aires. He began with a quotation from Nathaniel Hawthorne in 1881 referring in mythical terms to the power of the telegraph.

> By means of electricity, the world of matter has become a great nerve, vibrating thousands of miles in a breathless point of time... The round globe is a vast... brain, instinct with intelligence! (Gore, 1994a)

He argued that the development of the GII would be what would finally lead to the fulfillment of Hawthorne's vision. The prerequisite for sustainable development would be the building of this 'network of networks'. From it

> we will derive robust and sustainable economic progress, strong democracies, better solutions to global and local environmental challenges, improved health care, and – ultimately – a greater sense of shared stewardship of our small planet... (It) will help educate our children... It will be a means by which families and friends will transcend the barriers of time and distance... In a sense the GII will be a metaphor for democracy itself. Representative democracy does not work with an all powerful central government, arrogating all decisions to itself. That is why communism collapsed. (ibid.)

The European Union, through Commissioner Bangemann, took the information society to the global stage when it persuaded the G-7 group of countries at its summit in Naples in July 1994 to set up a Global Information Society Project. This was implemented through a G-7 ministerial conference on the Global Information Society in Brussels in February 1995. In the conference, Europe's 'information society' met the United States's 'information age'. Under the title 'A Shared Vision of Human Enrichment', the Chair's conclusions of this conference announced again:

> Progress in information technologies and communication is changing the way we live: how we work and do business, how we educate our children, study and do research, train ourselves, and how we are entertained. The information society is not only affecting the way people interact but it is also requiring the traditional

organisational structures to be more flexible, more participatory and more decentralised. A new revolution is carrying mankind forward into the Information Age... The rewards for all can be enticing. To succeed, governments must facilitate private initiatives and investments and ensure an appropriate framework aiming at stimulating private investment and usage for the benefit of all citizens.

In the United Kingdom, the concept of the information age had an earlier impact on policy than that of the information society. The first government document on these issues was a consultation paper titled 'Our Information Age' published by the Central Office of Information in 1997. In the foreword to the document, Prime Minister Tony Blair declared, in poetic terms:

The prize of this new information age is to engage our country fully in the ambition and opportunity which the digital revolution offers. The prize is there for the taking. We must stretch out our hands and grasp it. (Central Office of Information, 1998)

The 'information society' and the 'information age' eventually gave birth to the concept of the 'digital divide'. This is dealt with in the next section.

The digital divide

The global diffusion of the myths of the 'information society' or 'the information age' with their vision of a great technologically driven future which would bring untold benefits to all was accompanied by an increasing concern for the fact that some might be left behind. The G-7 group of countries sponsored a conference on the theme of 'Information Society and Development' (ISAD) which took place in South Africa in May 1996. The conference concluded that Information and Communication Technologies and Services have a potential to offer a significant contribution towards growth in all countries, but that a huge gap exists between the highly industrialised countries and the less-industrialised countries in terms of information infrastructure. It argued that developing countries were under-investing in ICT infrastructure and urged these countries to mobilise investment so that they can narrow the economic gap with industrialised countries.

Almost simultaneously international organisations, including the UN Economic Commission for Africa, UNESCO and the International Telecommunications Union, organised an 'Africa Information Society Initiative: An Action Framework to Build Africa's Information and Communication Infrastructure' in 1996. Africa was the first continent to undertake such a programme.

Whilst the concern for the international gap in access to ICTs had been highlighted by these initiatives, the concept of the 'digital divide' was born in the national context of the United States and subsequently adopted worldwide. It crystallised the fear that a gap was opening up in the United States between the 'information haves' and the 'information have-nots' which governments needed to help bridge. Its original use in 1996 referred to fears about the differential access to ICTs in different schools in America. This was allied to concerns that the United States might not be doing enough to develop skills for the 'information age'. Al Gore first used the concept at a White House ceremony in May 1996 when he said:

> we've tried, at the President's direction, to make certain that we don't have a gap between the information-haves and information-have nots. As part of our Empowerment Zone Initiatives we launched this cyber-Ed Truck, a book mobile for the digital age. It's rolling into communities, connecting schools in our poorest neighborhoods and paving over the *digital divide*. (emphasis added)

However, the term came into widespread usage as a result of a series of studies by the National Telecommunications and Information Administration (NTIA), a US Department of Commerce agency, undertaken at the behest of Vice-President Al Gore and published under the overall title of 'Falling through the Net'. The studies looked at the disparities in access to ICTs in the United States by geography, income, race and gender. All came to the same conclusion that the disparities were large and getting wider. The second of these reports, published in 1998 was titled 'Falling through the Net II: New Data on the Digital Divide'. The third report, published in 1999, was 'Falling through the Net: Defining the Digital Divide'. These reports received widespread media and political attention and were responsible for popularising the term.

The conclusions caused alarm in the Clinton administration and the development of policies for 'bridging the divide' became a high priority. However, at the same time, within the US administration the feeling developed that 'bridging the digital divide' could be associated with out-moded concerns with equality which Third Way Democrats had abandoned in favour of 'equal opportunity'. The White House therefore began to use what they considered to be the more fitting slogan of 'creating digital opportunity' for the marginalised. In February 2000, 'From Digital Divide to Digital Opportunity: the Clinton-Gore Agenda for Creating Digital Opportunity' was announced. This aimed to mobilise the private sector to help promote digital opportunity. The introduction to the initiative proclaimed that 'private sector competition and rapid technological change are powerful forces to...make Information Age tools available for more and more Americans'. At the same time Clinton announced US$2 billion over 10 years in tax credits to encourage private sector donation of computers, sponsorship of community technology centres and technology training for workers and smaller sums to help train technology teachers, promote

Community Technology Centres in low income communities and to help develop public-private partnerships to expand home access to computers. To launch the initiative Clinton undertook a whistle-stop tour of three deprived areas in the United States accompanied by senior executives of ICT companies. This 'digital divide' tour followed soon after the White House's top level forum of economists, business leaders and Wall Street analysts on the New Economy where the creation of digital opportunity had figured prominently. In a keynote address as part of a forum panel on 'Closing the Global Divide: Health, Education and Technology', Bill Gates enthused:

> These are amazing times. And I am proud and grateful to have the chance to be a part of the technology revolution at the heart of so much of the progress we are making. The scope of change – economic, social, and cultural – is awe-inspiring. Because technology has the power to make such a positive difference in people's lives, we have a simple obligation: spread it.

This particular way of trying to 'bridge the digital divide' and of creating digital opportunity has its ideological origins in the Third Way's triangulation method of developing partnerships between public and private organisations. However, it is also compatible with some strands of neo-conservative thinking. One of the gurus of the information age, Don Tapscott, writer of influential books such as *Paradigm Shift: The New Promise of Information Technology*, *The Digital Economy* and *Growing Up Digital* is credited as one of the early theoreticians of the digital divide. He explained his approach in an interview with neo-conservative talk-show host Geoff Metcalf, for Global Pathways, a web site which 'operates on the premise that the most powerful force in erasing the digital divide will be aggressive, private sector companies who provide products and services that meet the needs of the global market and make a profit' – the very approach being promoted by *Business Week* in the example given in the introduction to this chapter. Tapscott coined the term 'philanthropreneuring' for the strategy he was advocating. He argued that from a shrewd business perspective, companies should be able to find ways to benefit from their philanthropic activities in spreading IT across the divide through both tax advantages and brand enhancement which might well be greater than the benefits that they obtain from their advertising budgets.

    The 'digital opportunities' approach was carried into US foreign policy towards developing countries through the creation of the 'DOT-COM Alliance: Development in the Information Age' which was funded by US Agency for International Development (USAID) and sought to bring together public bodies, universities, NGOs and private companies with expertise in ICTs in order to provide support to USAID's efforts to 'bring the benefits of ICTs to under-served regions and populations'. USAID's strategy towards governments was encapsulated in DOT-GOV initiative of DOT-COM whose mission was to convince developing country governments to adopt 'policies that encourage private investment, competition and equitable regulation, leading to universal access and diverse service'. This included 'fostering privatization, competition and open networks, and universal service in telecommunications policy'.

At the same time as it was launched in the United States, the Clinton-Gore initiative was taken to the global level. The report by the World Bank's online magazine *Development Outreach* on the Davos World Economic Forum at the end of January 2000 was entitled 'From Digital Divide to Digital Opportunity: Business Leaders Report from Davos'. Clinton, Gates and many other leading executives of ICT companies were present. At the summit, John Chambers, President of Internet technology monopoly Cisco Systems, told his peers:

> We can change the life of every child who dreams of creating something new, but we must work together to create policies, practices, and opportunities to enable access for all.

In July 2000, the global information society and the digital divide became main themes of the G-8 summit in Okinawa, Japan. At the summit, the 'Okinawa Charter on Global Information Society' was adopted. The Charter stressed that

> (t)he essence of the IT-driven economic and social transformation is its power to help individuals and societies to use knowledge and ideas

and resolved

> to renew our commitment to the principle of inclusion: everyone, everywhere should be enabled to participate in and no one should be excluded from the benefits of the global information society.

In a section titled 'Seizing Digital Opportunities', the Charter highlighted

> the potential benefits of IT in spurring competition, promoting enhanced productivity, and creating and sustaining economic growth,

and called for

> (e)conomic and structural reforms to foster an environment of openness, efficiency, competition and innovation, supported by policies focusing on adaptable labour markets.

The G-8 created a Digital Opportunities Task (DOT) Force to address the problems of the digital divide. The DOT-Force in turn delegated, the following year, some of its work to Digital Opportunity Initiative, a public-private partnership made up of the United Nations Development Fund, the US Markle Foundation and the US-based global consultancy firm Accenture, which had re-incarnated from Andersen Consulting following Anderson's involvement in the Enron scandal. The Digital Opportunity Initiative adopted the slogan 'From Digital Divide to Digital Opportunities for Development'.

In December 2001, the UN General Assembly adopted resolution 56/183 which endorsed the holding of the World Summit on the Information Society, the first phase of which took place in Geneva in 2003 and to which I have alluded in the introduction.

Dealing with reality

The previous section detailed the way in which the concepts of the 'information society' and of the 'digital divide' have been consistently used to promote particular policies through the creation of dreams of a future techno-utopia in which all will participate, from which all will benefit, where the deepest human aspirations will be fulfilled. This makes use of the power of myth in mobilising human imagination to engender commitment to particular policies. It constitutes the creation of what Mosco (2005) has called the 'digital sublime'. The mythology is perpetuated through the incessant repetition of the mantra that the technology has a revolutionary potential to fulfil the deepest human aspirations, to create a world in which disease will be conquered, distance won't matter and communication will be both instantaneous and universal. At the same time the sense of community will be enhanced and new communities will be created unhindered by spatial or temporal barriers.

In order to achieve this, the myth has to create 'euphoric clarity' (Barthes, 1972, quoted in Mosco, 2005). It has to create a clear image, devoid of nuances and contradictions and devoid of politics. Insofar as politics involve struggle for resources between conflicting interests, they can have no place in myth creation for they would undermine the myth's effectiveness. However, from the point of view of the powerful and of those who stand to benefit from the policies that are being promoted, the use of de-politicised myths has the dual advantage that it hides the existence of winners and losers, whilst simultaneously capturing the imagination and mobilising people to actively support them.

Technological myths, as I have argued, convey to their creators the additional advantage that they hide human agency and particular political interests in promoting a specific direction of social change. Technology is consistently presented as the driver of the process: it is perceived as a rational, quasi-natural autonomous force independent of society and implicitly impervious to human agency. Accordingly, the phenomena which technology creates acquire the inevitability of natural events which human beings are powerless to affect. If some people are adversely affected, they must accept this as inevitable: resistance would be pointless and would achieve nothing. In any case, as the myth has created the mirage of progress towards a better society, such resistance can only be construed as standing in the way of desirable goals and inimical to the interests of the vast majority of humanity who stand to benefit from the technology-driven march to the Promised Land.

I have shown that the 'information society' or the 'information age' are presented by the powerful in ways that have all the characteristics of a technological myth. Myths are often couched in prophetic and poetic language as in the case of Tony Blair urging us to stretch out our hands to grasp the prize that the digital revolution offers. When, in similarly poetic terms, the President of Cisco Systems claims that we can change the life of every child who dreams of creating something new, but that in order to do so we have to work together to enable all to have access to the technology, he appeals for support for policies that will benefit his company in terms that appear altruistic whilst creating an image that collective solidarity is the means to achieve this. A similar appeal to higher human aspirations is made when the Swiss President claims that the World Summit on the Information Society has coined the new term 'digital solidarity'.

I have also shown that the inevitability and benefits of the information society are systematically associated with the expansion of the market and the dismantling of government control and regulation. This ideology embraces the claim that the full benefits of the technology can only be enjoyed by opening up markets. 'Building the Information Society for All', a positive slogan implying social solidarity adopted by the European Union, becomes simultaneously implicitly associated with de-regulating markets. The market said to be in most need of de-regulation is the labour market. Its regulation, whether by governments or as the product of effective collective action by trade unions, is presented as one of the greatest obstacles to the development of the information society.

However, the reality of disintegrating social solidarity and of increasing inequality contradicts the myth. We are told that there is a revolution towards a techno-utopia which we should all welcome wholeheartedly, but numerous studies demonstrate that the *economic* divide between the haves and have-nots has widened. This is true of the divide between countries (Cornia, 2004).

It is also true of the divide within countries. Within the United States, the country that has travelled fastest and furthest towards the 'information society', inequality has risen at the highest rates in the Western world in the course of this revolution both in terms of income and wealth. By some measures the increase in inequality is not just a relative phenomenon. It is not just that the real incomes of the rich are rising at a faster rate than those of the poor: the real incomes of the rich have risen whilst those of the poor have declined. The biggest winners have been the super-rich (Cassidy, 1997; Krugman, 2002; Piketty and Saez, 2004).

This huge rise in inequality created a considerable degree of political and academic concern in the United States which was difficult to escape. The Clinton administration addressed this issue in the *Economic Report to the President* of 1994. The report recognised the rising inequality and called it 'a threat to the social fabric'. Included in the list of factors responsible for the rise in inequality were the spread of new technologies, diminished union strength and a falling real minimum wage. However, despite the fact that rising inequality was being officially ascribed partly to the spread of new technologies, the administration chose to focus the problem around the concept of the 'digital divide' which it then successfully transferred to the global stage. The implied medicine was one of 'hair of the dog'. Inequality could be resolved by accelerating the spread of the new technologies which were partly causing it. It was only necessary to ensure that those who had been left behind could be brought into the fold and included.

There is, however, no evidence that ICTs will make any contribution to closing the socio-economic divide. If anything, as I have shown, the evidence points in the opposite direction: it is unlikely that creating 'digital opportunity' would do anything to ameliorate inequality within countries. Even if it did, the problem is that, as one of the architects of the Clinton policy has since admitted, the excluded could only be included through government action that involves serious investment (Reich, 1999). 'Philanthropreneurism' will never be more than a means for corporations to improve their image by pretending to be addressing the problems of the dispossessed. It is also unlikely that the villagers of rural India or the women of North East Brazil as customers will make much contribution to Hewlett Packard's balance sheet in the foreseeable future. However, serious government investment requires serious government money, and this is incompatible with the strategy of reducing taxation and embracing the market which has characterised the Third Way.

Conclusions

Technological determinism perceives the relationship between technology and social change as one in which technology is a quasi-natural force driving social change in a way that is impervious to human choice and human action. This chapter has contended that technological determinism is an ideological weapon favouring the powerful and seeking to hide the fact that there are winners and losers from change. The concept of the 'information society' is presented by politicians, business executives and the media in technologically deterministic terms. The terms used also give a mythical status to this concept, creating visions of a techno-utopia. This is designed to mobilise support for, and undermine opposition to, the changes associated with the policies being promoted. Approaching the problem of socio-economic inequality through the concept of the 'digital divide' perpetuates mystification and prevents real problems from being addressed.

'Building the information society for all', 'bridging the digital divides' or 'creating digital opportunity' are therefore inappropriate slogans for those concerned with rising inequalities in society. 'Digital solidarity' is not an appropriate way to address concerns with increasing individualisation.

The issue is building a *society* for all. This can only be done by 'bridging the *socio-economic* divide'. Such bridges cannot be built without challenging wealth and privilege. Social solidarity cannot be constructed by embracing the very mechanisms which promote the increasingly individualised consumption and economic greed which the continuous expansion of markets requires of society.

Notes

1. Influential academic works in the development of the concept of the 'information society' include Bell, D. (1973) and Castells, M. (1996). For a critical discussion of the theories see, for instance, Lyon, D. (1988) and Webster, F. (1995). Barbrook (2007) highlights the technologically determinist and mythical nature of the concept and traces its origin to Marshall McLuhan's 1964 book *Understanding Media*.
2. The term 'multimedia' is widely interpreted in technological terms as the integrated technology that results from the convergence of the previously separate technologies of sound, text and image. This convergence has been made possible by the digitisation of all forms of information.

3. Inequalities in the Globalised Knowledge-Based Economy

RICHARD SHARPE

Introduction

Information, or content as many publishers call it, increasingly appears to be a free good. It seems to be owned by anybody and everybody. Anybody, it seems can and is encouraged to contribute to the base of information in this knowledge-based economy. The old order of property seems to have been overthrown. Yet underlining the knowledge-based economy are the same rules of property in the form of intellectual property.

Free services and free information

The world's most used web sites at the end of October 2007 were, in order of use, Yahoo, Google, MSN, YouTube, Windows Live, MySpace, Facebook, Wikipedia, Orkut and Hi5 (Alexa). The list is led by three Web 1.0 search engines, and soon followed by six Web 2.0 community sites and Windows Live. All 10 are free to the user. Users seem to have free access to the knowledge of the world within seconds delivered to them at their home, office, classroom, library or on the move.

It is an experience which gives them the personal power to access and transform knowledge far in excess of the power of previous generations. They are part of a new society, Castells (1996) argues, 'based on knowledge, organised around networks, and partly made up of flows'. This new society stretches beyond the old boundaries of society; the user could be anywhere with telecommunications links and still be able to access the same knowledge and contribute to the knowledge society. It seems to be a knowledge-based economy. Such is the current myth.

There seem to be none of the old discriminations which dominated the industrial and previous societies. Men and women can use the tools of the networked society equally. Low prices mean that wealth is a minor barrier. The Internet is blind to the colour of the user. And the location is increasingly becoming irrelevant. As Frances Cairncross (1997) argues, in the *Death of Distance*, 'no longer will location be key to most business decisions'. So the old inequalities of class, gender, location and race have been swept away or are being eroded.

It seems as if knowledge has replaced property: the brain has replaced the factory; manipulation of symbols replaces the manipulation of atoms.

We can search with Yahoo and Google: search other web pages, images, blogs, academic papers, maps and buy products. We can even translate material for free. We can contribute to communities of friends, family and colleagues using the Web 2.0 sites. It is, after all, a post-industrial society, a phrase coined by Daniel Bell in 1973. A few own factories or other parts of industry. But all humans have knowledge. We can all think. This way the post-industrial society seems to be a release from the limitations of yesterday. Bell is well aware of the differences between knowledge and property. He says that the social forms of managerial capitalism are likely to remain for a long time. They are 'the corporate business enterprise, private decision on investment, the differential privileges based on control of property' (Bell, 1973: 372).

This displacement of property for knowledge is taken up consciously or unconsciously by other promoters of the technologies which came to provide the Google users their experience. An early one was James Martin (1978) who popularised the notion, the myth, of the 'wired society'.

So Bell and others provide the overarching structure while Martin and others fill in the panels with pictures of the digital sublime. They predict the future anticipating the impact before the technologies are fully formed which will have the effects they predict. They create and amplify the myth.

Property creates boundaries

Property immediately assumes some boundary between the ownership of one person or group of people and the non-ownership of others. Property immediately introduces an inequality in social relations: the haves and the have-nots; the owners of property and the owners of less or no property; yours and mine; ours and theirs. The rich and the poor, with grades between them. Property is not evenly distributed.

The property central to the networked society is the property of knowledge, called first industrial and later intellectual property (IP). IP differs from physical property which was the dominant form of property in previous phases of the capitalist economy. Pre-capitalist societies developed the notion of property, of individual and collective ownership, which included some and excluded others. This tribe and not that 'owned' this land. This family and not that had the land by the river. This individual and not that owned the forest, this beast, that dwelling.

The modern notions of IP were formed by white, property-owning males in the metropolitan centres of Western Europe, particularly England, from the late seventeenth century onwards. Modern notions of IP in the UK cover

- The expression of ideas – copyright;
- The development of new methods – patents;
- Marks for trade – trademarks;
- Secret information – confidentiality;
- The goodwill of a trader – passing off.

These white, property-owning males in metropolitan centres from the late seventeenth century onwards who developed the modern notions of IP did so in order, partly, to defend the work of fellow white, property-owning males in the metropolitan centres who were then developing information systems in the Age of Reason and Revolution, as Headrick (2000) describes it. They were part of the overthrow of absolutist notions and control exercised by monarchy and church. These innovating men were producing the modern systems of classification, nomenclature and measurement; they used statistics and numbers to describe the world, and they displayed information visually.

At the end of the seventeenth and the start of the eighteenth century, the absolutist methods of controlling printing, the main means of distributing ideas long distance, collapsed in England. A Royal Decree of 1586 had concentrated printing in London, apart from the Universities of Oxford and Cambridge, and had limited the number of presses to 22. York was added in 1662. In 1695, in the new spirit of more liberal trade after the collapse of absolutism in England, the Royal Decree limiting printing presses – the absolutist control of copying – lapsed. This led to a rapid expansion of printing, and of, in the eyes of the writers and original publisher, piracy. By 1725 there were presses in 14 other cities outside London, Oxford and Cambridge (Febvre and Martin, 1976: 191–192).

As absolutism declined, the waning power of the Crown to control publishing left the market open to what the London-based publishers saw as pirate copiers. They petitioned Parliament to protect what they called their 'literary property' in 1707. They eventually won their argument and Parliament enacted the copyright act in 1709. 'The Act was a booksellers' [publishers'] act not an authors' act, and it seemed to represent precisely what the copy-owners were seeking', despite the fact that it was called an 'Act for the encouragement of learning' (Feather, 1988: 75). The property right to copy a book you have purchased was separated from the property right to own the book itself. Statute law overriding common practice was the tool this group of property owners used to enforce this distinction between the right to own and the right to copy. Some were women who had inherited their copyrights from their fathers. The Statute of Anne 1709 granted copyright to the copyowner for 14 years. If the copyowner was still alive after that, it was granted for another 14 years. Others could buy the book, could read it and could give it to another. But they should not copy it. Content was separated from form. The value was in the content, not only in the form.

Simultaneously to the popularity of almanacs and religious publications along with the rise of the expression of ideas in the age of Reason and Revolution, the modern English novel was taking form. The creative output of these novel writers was supplemented by the expansion of English poetry, the theatre and, later, the rise of periodical magazines and newspapers.

This expansion of publishing of the printed word was achieved without any change to the basic technologies of printing. The Gutenberg press had been slightly adapted in the 1620s with the aid of a counterweight which enabled printing of up to 150 sheets an hour on a single press. This speed of output remained the same until the development of the steam press in the 1810s.

Notion of intellectual property established

Now that the notion of IP was established and new IP being created, legislators and judges, all white, property-owning males in metropolitan centres, extended the notion of copyright in three directions: to cover other works apart from books, to extend the time of protection, and to cover other locations. In England they added engravings, lithographs and prints in 1734; sculptures in 1798; drama in 1833; and music in 1882. And on into today with software, films, the content of web pages and Blogs, broadcasts, typographical arrangements, sound recordings, architecture, cable casts and choreographic works.

Gradually, and inexorably, the time of protection, the time when only the copyright owner could copy, adapt or translate the work, was lengthened beyond that granted in the Queen Anne stature of 1709 by legislators to the benefit of copyright owners. By 1908, it was 50 years from creation. By the 1990s it was 70 years from the death of the author, if the author was known, or of its creation. Consistently those with interests in owning copyright and gaining from its ownership were able to persuade legislators to extend their monopoly rights to copy. And when they did win an extension, it was always against the arguments of those who argued for the common ground, the public domain.

IP owners also spread the notion of copyright to other locations when their reach extended. In the 1750s, England became the dominant power in the western European metropolis as a result of military victories in North America, India and through the confinement of French naval power. British Empire colonial administrators and judges took the same IP notions to the growing Empire and implanted them there, with London as the focus of publishing.

Even when parts of this Empire won their independence through struggle they often adopted the same form of copyright laws. The Statute of Anne was incorporated into the US constitution in 1790. It was extended from books to also cover maps and charts. These were important parts of IP because they showed the way for expanding physical property beyond the limited eastern coast of the United States. Noah Webster, a successful author of educational textbooks and later of the first definitive American Dictionary, was a prime mover in getting the new Congress to adopt, and later extend, the period of monopoly of copyright. Samuels (2002) calls him 'The father of American copyright'. An appropriate gender-specific title, another white, male, property owner stakes his claim to his intellectual product.

As IP generation got under way in the United States, as it became IP-rich, the Republic's legislators added other types of media in addition to books, maps and charts. Samuels (2002) illustratively chronicles this growth to include sheet music, piano rolls, phonograph recordings, public performances for profit of music, radio transmissions and films. Samuels shows that copyright did not stand still, but was pushed and expanded by the owners of IP. The basic notions of copyright withstood the pressures and changed economics of many 'technological revolutions'. For example, 'the existing copyright law – together with some aggressive self-help by the composers and publishers themselves – was enough to take care of the radio revolution'.

International coordination of intellectual property

Eventually American and other property owners realised that they should respect the IP of property owners in other nations as the means of reproduction spread and markets internationalised in the 1880s. They formed a common cause with other IP-rich people. They handled this spread of the notion of copyright by bilateral treaties. Eventually the white, male property owners realised that a more stable form of protection from others and between themselves could be forged by a more permanent agreement between them. Metropolitan nations and empires decided in the mid-1880s to collaborate in order to grant the subjects of each of them the same rights of property ownership and they forged the Bern Copyright Convention of 1886. The Bern Convention was more than just the sum total of countries signing the convention. It was a cooperative organisation of nation states operating on behalf of their IP owners in common cause to extend IP rights. It was a supra-national authority, above the states which signed up: supra-national because the notion of IP was not sufficiently established between these nationals signing it.

Copyright is a defence of IP by the IP-rich against the IP-poor. Traditional media companies use it to defend their IP against the Web 2.0 upstarts. For example, in March 2007, Viacom hit YouTube with a $1 billion suit for copyright infringement for allegedly carrying over 160,000 clips from Viacom TV programmes.

Industrial property became important for profit generation from the middle eighteenth century. Property was to be generated based on the development of methods of manufacture implemented in the work of the industrial worker. As these were also property they needed protection. So property owners developed the modern notion of patents. One of the first to seek this protection was Richard Arkwright, inventor of the water frame for spinning cotton. Patents then were not strong enough to protect his IP (Black and Porter, 1996). They picked up an early-modern notion of restricting the range of trade, and of the import and export of skills granted by absolutist monarchs in letters patent. Patent meant open, unsealed letters which could be shown to anybody who asked to see them. The Crown had used these patent letters in the early-modern period to circumvent the control of manufacture and trade by guilds.

Patents are monopolies on an invention or method which is granted to the first person or group of people to apply for them, as long as they have not taken only from common methods or copy the invention of others. This notion of patents has developed from inventions to, among other things, provide protection for the IP of companies trying to exploit the biodiversity of the developing world (Senker, 2000: 211).

The English Parliament passed a Statute of Monopolies under James I in 1624 which granted inventors 14 years to exploit an invention: two apprenticeships of seven years in order to train others in the invention. The early inventors of the industrial revolution used this method to try to protect their property when the mobility of labour became more widespread in the eighteenth century. Secrecy was the best method of protecting inventions or even discoveries. But mobile labour began to erode this method of secrecy as workers could travel with their knowledge from one employer to another or even set up on their own. In the 1750s, 10 patents were granted to British inventors. As the pace of the industrial revolution grew, more and more patents were granted. In the 1760s, 20 were granted; by the 1810s it was 110 a year; by the 1840s it was 458 a year (Cornish, 1989). The 1851

Great Exhibition's focus on industrial innovation led the new industrial property owners to press for more protection for their property which they achieved in a reform of the UK patent laws in 1852. The fee for registering a patent was slashed to £25 from £300, a 92 per cent cut (to £1,596 in money terms equivalent at 2006 prices from £15,917 (Officer, 2007). The number of applicants soon doubled: to 891 in 1851; to 2,113 in 1854; to 4,000 or more a year in the 1870s; and to over 9,000 a year in 1883 (Cornish, 1989).

All of the top 10 web sites use patents to protect their IP. Google, for example, holds over $155 million worth of patents and developed technology in acquisitions alone in its asset books (See for example BBC News, 1 November 2005).

As the European metropolitan property owners in countries outside Britain realised the importance of this process of industrial innovation, and the need to defend the knowledge it was based on, they also adopted patent laws. The Austro-Hungarian Empire in 1873, for example, invited other countries to exhibit their innovations at a great exhibition only to find that few would exhibit because the Empire had no patent laws. Anybody could see the invention at an exhibition and copy it. So the Austro-Hungarian Empire passed its own special patent laws to make international exhibitors feel that their property would not be copied (Cornish, 1989). Germany followed suit with a patent law in 1874. This form of IP spread beyond Europe. For example, Japan emerged from its period of isolation in the 1850s and adopted a patent system in the 1880s.

The IP-rich, also needed to extend the recognition of their ownership rights in other metropolitan areas if their own property was to be defended. The Austro-Hungarian authorities called together an international conference to discuss the issues in 1873. More successfully, a decade later, the metropolitan property owners gathered in Paris to grant rights to each other so that foreigners could now apply for patents in their countries. This international cooperation mirrors the Berne Convention for copyright and, as with copyright, the monopoly in the patent was extended, this time to 20 years from 14. Again, wherever the metropolitan powers extended their empires, so they extended their patent laws.

The two pillars of IP – copyright and patents – were then well established internationally by white men of property from the metropolitan areas by the 1870s. In addition to copyrights and patents, the English legal system developed the law of breach of confidence from the middle of the nineteenth century. This was firmed up by two cases brought by IP owners: one brought by Prince Albert and another by a patent medicines vendor (Bainbridge, 1999).

The trademark laws were also established in 1875 to protect the property of a mark of trade or service. The first trademark in the UK granted in 1876 was for the Bass brewers triangle. Brewing had become a nation-wide industrial endeavour replacing local breweries (Cornish, 1989).

From the 1840s as competitive trade in consumer goods began to grow, English property owners developed another part of IP law to complete the picture: passing off (Torremans and Holyoak, 1998). 'Nobody has the right to represent his goods as the goods of somebody else' is the classic definition (Cornish, 1989).

Breach of confidence covers the early development of ideas before they are expressed. Copyright covers the expressions of the ideas. Patents cover any development of new processes. Trademarks cover the signs used to attract customers to the goods or services which have been developed. And passing off provides a wider defence of IP when the goods are traded.

By the time electronic, digital, stored-instruction computers were developed and, later, the Internet, this framework of IP law was firmly established from its origins to protect the interests of white men of property in metropolitan areas. The development of technology in subsequent centuries did not mean that this framework of IP defence had to be radically reformed. It need only be extended to cover the new media. As Whinston, Stahl and Choi argue (1997: 181) about copyright, 'The invention of printing presses, photocopiers, and now digital copying technologies have periodically brought the issue [of illegal copies] to the forefront. But the market environment has not changed significantly, and the digital marketplace does not present any new issues that demand a complete revision in IP laws as some have argued.' No nation has attempted a complete revision of IP laws in the light of the new pressures on property in the knowledge economy. Instead IP owners have extended the basic notions in time, space and content. For example, patents have been extended in the United States to cover software which was seen until the 1990s as a literary work and only defended through copyright.

Let us now turn to the network, and its use as a myth.

The network used as a myth

Even in the notion of the modern Internet, the network itself is mythologised. The network is presented as a network of equal nodes. The model of the Internet as a packet-switched network without a centrally controlling host or hosts is transferred to the power relations of nodes and their users, as if all nodes are equally free and powerful. But equality of all nodes and access to all information is not the reality of the network behind the search engine or Web 2.0 site on which that experience rests. The myth says that the experience of the consumer is the reality: an easy experience to encapsulate in a myth and so to subdue the confusion and complexity of reality. But here the myth distorts the reality.

All nodes are not equal. The real IP is hidden behind nodes which will not allow access to the rest of those without the power to access this important property of knowledge. The vast volume of knowledge in the public domain, indeed, gives the individual user the impression that all nodes are equal, that the search engine is searching the sum total of human knowledge available in a digital form. This is far from the reality.

These private owners of IP constantly fight with hackers and others to create more security for their information precisely because they want both to use this technology and also to preserve the value of their private property. They want the benefits of openness while defending their IP riches.

Knowledge is often only of value if it is held in private, if access to it is controlled and restricted. Property value is based on exclusive, monopoly, access, whether it be physical property or IP. Share those rights to access or use and the value tumbles: in other words, when it comes to the value of property, the tighter the monopoly, the more the value. The more IP-riches, the higher the value.

The IP-rich coordinate globally

IP-rich nations, pursuing yet more IP-value, launched another round of coordination of the IP-rich in the later part of the twentieth century. The increasing internationalisation of markets prompted another round of attempts to coordinate IP defence a century after the initiatives of the 1880s. The World IP Organisation (WIPO) was founded in 1995 as a UN-sponsored organisation aimed at harmonising IP laws across the globe. It is another supra-national organisation looking after the interests of the IP-rich beyond national boundaries. WIPO had 181 member countries in early 2005. It operates by calling together international conferences at the level of diplomats who then thrash out a draft treaty. When this draft treaty is ratified by enough member nations, it becomes WIPO 'policy' and has to be incorporated in the national laws of all member states if the nation wishes to remain a member of WIPO. This process is given particular clout by being linked to membership of the World Trade Organisation (WTO), the pact of 148 nations which operates a low tariff/no-tariff trade policy towards each member.

If your nation wants to gain the benefits, as the leaders of member countries see it, of low or no-tariffs prompting higher volumes of world trade as a member of the WTO, then you have to adopt the notions of IP generated by the white, property-owning males in metropolitan centres. WIPO not only has its treaties but also has what it calls 'soft laws'; these are recommendations to member nations. So far, these soft laws include recommendations on trademark licensing, well-known marks, marks and other industrial property, and the property rights in signs on the Internet.

The leaders of many countries of different political histories want the benefit of WTO membership and are, therefore, obliged to adopt the WIPO

IP framework. Even the Chinese copyright law bears a remarkable likeness to the WIPO copyright treaty, however it is implemented. Communist countries, or those led by Communist parties depending on how you define China, are willing to adopt capitalist notions of IP in order to preserve and expand their role in world trade.

There is a clear imbalance in the ownership of IP, as one would expect in the distribution of any form of property in capitalism. We have the IP-rich and the IP-poor, to take the notion of copyright rich and copyright poor from Vaidhyanathan (2001). WIPO figures for the number of patents granted in the country to residents and non-residents are clear indicators of which countries are generating patents internally and which are having their citizens register patents in other countries so as to take advantage of the international patent coordination regulations.

The United States is one of the few countries with more residents applying for patents than non-residents, according to the annual league table of patents issued published by WIPO. A total of 86,976 were issued to residents in 2002; 80,358 to non-residents. Many countries have no patents granted to residents in that year, but several to non-residents. These countries include, just from A to G: Albania, Azerbaijan, Belize, Bosnia, Botswana, Costa Rica, Dominican Republic, Gambia, Ghana and Grenada. Even relatively prosperous countries in Europe with significant amounts of innovation cannot generate the volume of patents granted by residents close to those granted to non-residents: among them Austria with 1,581 to residents and 18,809 to non-residents; and Finland with 183 to residents and 11,406 to non-residents.

De facto standards defended by IP

De facto standards are central to the Internet and the workings of the knowledge-based economy. No de facto standards and no communications. De jure standards such as the OSI communications standard and the Ada programming language standard often become the backwaters of development compared with the de facto standards of TCP/IP and C, to mention but two. De facto standards mean some degree of monopoly, if they are to be real de facto standards. Talking about software standards alone Whinston, Stahl and Choi argue, 'When a product becomes a de facto standard and is protected by copyright, its producer indeed employs a monopoly market power. In many cases, such a monopoly is encouraged to minimise duplicative costs of having competing standards. That monopoly, however, is often regulated in exchange for its monopoly power. In today's Internet environment, where regulation is rejected from all sides, a dominant firm will have an unrestricted market power after its product becomes the standard. Copyrights for its product in turn protect its monopoly position unless other firms are allowed to licence it' (1997: 206).

In other words, the perpetuated myth of the open and free Internet serves to create the conditions of monopoly for selective IP owners.

There is another important aspect of IP and monopoly which is often overlooked: the search engines of the search sites. Yahoo and Google, for example, do not publish the criteria by which the sites are selected for listing when the search is completed. Few searchers get beyond the first page. Getting onto the first page or, even better, into the top three on a search is the Nirvana of web site owners. It is called Search Engine Optimisation (SEO). As the search engine sites won't tell the world how they do it, SEO becomes a black art. In effect, by making their selection criteria a piece of their IP and defending it, the search sites have created a mini industry playing much the same game as the Kremlin Watchers in the Cold War.

Antitrust laws are clearly hardly a defence against these types of monopolies: de facto standards monopolies and monopolies in search algorithms. Most monopoly laws were devised to stop the deliberate and predatory 'robber baron' activities of industrial capitalism, the target of the late nineteenth-century American populist movement. These laws require that those judged to have a monopoly or strong market control are proven to have taken some monopolistic action, such as price fixing, ruling out some customer, deliberately delaying innovation or bundling. But as Whinston, Stahl and Choi (1997) show above, a monopoly owner need not take any predatory action in the knowledge economy to be a monopolist. All he needs to do is to become the de facto standard and protect that market share with the legal framework of IP laws to continue to create inequality.

Monopolies are not an attack on free markets, but the outcome of a market for property in which property is protected by IP laws. Critics of the current trends of IP laws rail at this. Lessig (2002) argues that the Internet provides a common neutral platform on which ideas can flourish. Perelman (2002) is particularly critical of the current development of IP as he sees IP rights as a means to confiscate creativity. Perelman argues 'the most important of modern technologies depend upon creativity from a time that predates the recent revolution in IP rights. Public investment funded most of this fundamental research. In addition to public science, the public at large has built the educational system that created the infrastructure that made modern science possible. Nonetheless, private interests confiscate most of the fruits of this public investment, claiming their reward on the grounds of their ownership of IP' (2002: 210).

This misses the point. There has not been a revolution in IP rights, just a steady extension of their coverage and application by media and location by those in political power on behalf of the IP-rich. Perelman, Vaidhyanathan and Samuels talk separately of a revolution, of a fundamental change in IP; in their US tradition they often point to the limited monopolies and coverage of media granted by the 'founding fathers' of the US republic for IP monopoly. They adopt a minimalist interpretation of the US constitution: what the founding fathers meant was good enough for the history of the republic. Instead, the history of IP is a steady imposition of property rights by those who own property. There need be no major revision of the laws of IP ownership: just a steady expansion as more is generated. Samuels provides the evidence for this argument throughout, while all the time claiming there is some fundamental change afoot.

Perelman and others argue that public funds often kick off the creation of IP, and therefore it should be in the public domain. Public investment in a capitalist society is often made for private gain. And those who made this investment have more often than not been able to appeal to the legislature and executive, often populated by male property owners, that the rights of private property should be defended by the military or extended.

The knowledge workers in capitalism lose their IP

But, it can be argued, IP is different from physical property. We do not have to occupy the ideas of others in order to use them creatively. If I copy an article, I have not taken it away from the owner. The more the product of intellectual endeavour is freely available, the more creativity will flourish. Yet creativity is, after all, only the output of the knowledge worker. Why should the knowledge worker within capitalism have any right to own or control their own output any more than other workers? After all, under copyright laws, the product of the worker remains the property of their employer, including its copyright. Software developers within the tradition of handicraft production had the notion that the software they generated was theirs. This relationship of the producer to their product reflected the control labour had over the production process of software and, hence, management's lack of control. In the 1970s, textbooks for the managers of programmers made it clear that programmers 'must be made to understand' that they are working on a corporate asset, not their own property (CSE, 1980: 34). Programmers were producing a public asset, that is an asset for the company, not a private asset which they owned. A librarian's role was devised to capture this public asset. 'This identification of all program data and computer runs as public assets, not private property, is a key principle in team operations' (Duncan, 1981: 185). The owners of capital need to split the ownership of the product from the producer: the powers of capital have set capital in motion to generate more profits, not to let the producer own the product.

Employers lobby to make sure this ownership of the product does not happen when IP legislation goes through legislatures. It may be only because both of these authors, Lessig and Perelman, and indeed many authors, are knowledge workers in academic institutions that they raise this issue of the ownership of the product by the producer in twenty-first-century capitalism with any seriousness. It has mostly been destroyed elsewhere. Indeed, the rise of the .com entrepreneur economy in the 1980s can be seen as confirmation of this: IP workers had to become company owners themselves to own the products of their own labour. If they remained employees they would not own the intellectual products, as it is made clear in the copyright acts. Both Perelman and Lessig work in the knowledge factories of the American academia, called the diploma mills by Noble (2001/2004). This point of departure by Perelman and Lessig, among others, is the same as that of Stallman (1998). The imposition of IP within the academic world where Stallman developed software meant that 'a cooperating community was forbidden'. Private IP became dominant by those seeking to benefit from their control of it.

The work of Stallman and others who have developed the free and open source software movement shows that there are continued battles of resistance to the domination of IP owners. The fact that they have to fight such a battle shows the power which IP owners have. The fact they seem to be fighting a limited battle against strong odds with scattered victories adds power to my argument that this is a fundamental issue of property, not of the particular nature of networks, of the special values of software development or of the creative world.

It is essential for the continuation of capitalist property ownership that the worker who produces physical objects should not own those objects. The objects are made in order to exploit their qualities as exchange value, to gain the profit of trade. As the human mind becomes a direct productive force, so the products of the human mind, IP, also become valuable for their exchange value. These products are owned by employers: those who set the productive cycle in motion by the expenditure of their capital. Only if these intellectual products are owned by those who set the productive cycle in motion, can those who make the initial capital investment exchange them for their benefit, and not for the benefit of the creators. No wonder the owners of IP fight so hard to extend their domain. The free distribution of knowledge attacks the very foundation of their ability to generate profits.

Generating profits from 'free' information

The users of the top 10 sites, with whom I started, seem to have a free service. But the service is not free: it is being subsidised by those who want access to those users. And the users are paying through a tariff on the goods they already buy. The users have paid through the goods and services they have or will pay for because the vendors of those goods pay for advertising to get the attention of those users. Yahoo is seen by 27.6 per cent of all Internet users a day, says Alexa. Google by 25.5 per cent. Of the $4.2 billion revenues of Google in the third quarter of 2007, 99 per cent was paid for in advertising by vendors who want to attract the Google user. The top sites have the same business model as commercial TV: the consumer pays through the goods they buy and receives a marginal service of entertainment/information search seemingly for free for the individual but actually paid for by the collective of consumers. And further, Yahoo, Google and others charge to place certain sources of information higher than others. This part of the search engine business model is like entering a library and seeing the books that have been supplied by publishers willing to pay more high-lighted, from the normal classification of alphabetical for fiction and Dewey Decimal for non-fiction and made much more prominent than those who do not pay. In other words, search engine sites sift their information according to strict commercial criteria.

Yet further, the web user has paid, directly or indirectly, for access by paying for the IP in their PC for operating systems and applications, and in the servers of the ISP. In the UK, leisure services are the second highest item in household expenditure (ONS1 and ONS2). In 2005/2006, they were 14 per cent of average household expenditure. Internet services are, indeed, only a small part of this but, apart from sport, they are the fastest growing. Internet subscriptions rose 800 per cent from 1999/2000 to 2005/2006 compared with total family expenditure rising just 5.9 per cent. And this is all before any proprietary sources of information are accessed and paid for.

Conclusions

The web user has the false impression of the consumption of free goods and services in a networked world of equal power. This impression of free consumption fuels the myth of free and open access to information, masking the role of property ownership.

This powerful myth of free and open access does not reveal the inequalities of property ownership, which will exist as long as the knowledge-based economy exists within capitalism.

4. Old Wine in New Bottles: How New Technology Recycles Old Myths

BERNARD KAHANE

Introduction

In this chapter, we first consider how successive new technologies (Information and Communication Technologies (ICTs), biotechnologies and nanotechnologies) seem to occur as recurrent waves which shape society. Such waves are accompanied by ideas about them which include promises that have the potential to create momentum and/or social disruption. We then identify the main characteristics of these waves. In the second part of the chapter, we summarise some of the positive mythical stories disseminated by promoters of technology and those which counter them, mobilised by opponents who face new technology waves as they emerge. We emphasise that opposition to new technology may be legitimate and conclude that, whatever the new technology, there are common patterns in the myths which accompany new technology waves.

Common patterns in new technology waves

The present era is shaped by successive waves of new technologies which disturb the existing order of the world. This has important consequences for those who produce science and technology as well as those who may benefit or suffer from their impact. Information and telecommunication technologies express themselves as new ways to tap into the potential of the digital world through manipulation at the level of the *bit*. Biotechnologies express themselves as new ways to tap into the potential of the living world through manipulation at the *molecular* level. Nanotechnologies (the new game in town which builds to some extent and impacts upon ICTs and biotechnologies) express themselves as new ways to tap into the potential of matter through manipulation and properties at the level of *atoms*.

Such disruptive processes act through three components: science and technology in which research organisations and practitioners have to change their practices to accommodate new technologies; industry and business, in which companies may either foster new technologies or experience a competence destroying process; markets and societies in which consumers and users may develop new practices or even components that can have significant implications for the evolution of the new technologies. (Thomke and Von Hippel, 2002) As Schumpeter pointed out, new technologies are at the core of creative destruction, producing variability while testing existing position, assets and potentials of companies, workers, consumers and users. Through this production and selection capacity, new technology waves exert tremendous impact on the evolution of society (Schumpeter, 1989).

New technology waves have their origins in complex mixtures of scientific and technological knowledge (mathematics, physics, chemistry, biology, computing, electronics etc.), and progressively gain strength and momentum under the influence, participation and support of interested actors (researchers, entrepreneurs, companies, users, policy-makers, media) before delivering their power onto the shore of wider society. Difficult to perceive at first, in time they become too big to be missed. Interested parties (stakeholders) may wish to avoid or catch and surf these waves, crawling to escape or reach them in a process that is often easier before the waves gain full momentum.

Four dimensions characterise new technology waves: first, they disrupt the prevailing order. Under the influence of new technologies, a reshuffling process occurs. Incumbent companies are challenged by newcomers while industries and markets are drastically changed. For example, our way of taking and processing pictures has been radically transformed pushing established companies such as Polaroid and Kodak to adapt. Travel agents and their employees are threatened by consumers' ability to book their travel online while new actors emerge. Biotechnology firms such as Amgen and Genentech have been able to compete with big pharmaceuticals and to introduce blockbuster pharmaceutical products in less than 10 years. In a short period, software and hardware companies such as Microsoft and Intel have become world leaders, forcing established firms such as IBM to reinvent themselves or face extinction.

Second, new technology waves involve not only new products and new processes but also, in the absence of precise explanations of the rationale behind known phenomena or practices, have promoted the capacity to extend what people previously did. In relation to new technologies, humans are often like Monsieur Jourdain who wrote prose without knowing it (Molière, 2005). For example, biotechnology gives access to gene manipulation and helps to explain how genes work and are regulated. For centuries, people grafted and created hybrid plants, but they now get access to the internal process and can predict what will happen if a specific gene is inserted in a specific location. In the Middle Ages, craftsmen included gold or silver particles in glass so as to obtain specific colours for decoration of churches and other buildings' through a process called sublimation (Chang, 2005). The colour produced depended on the particles' various shapes and sizes, but craftsmen could not understand why a specific particle produced one colour rather than another. It was only at the end of the twentieth century with the introduction of nanotechnology that the rationale began to be understood, creating opportunities for further developments. Thus, by virtue of the theory which they incorporate, new technologies can offer ways of understanding and extending what was previously a matter of practice and observation. They can also involve the development and use of new tools and instruments which create possibilities of new ways of working. Examples of this are many: genomic sequencing involved the development of automatic sequencing machines; electronic chip design tools and the vast amount of instrumentation involved in producing and checking chips reliably created new potential for information and communications technology; nanotechnology development required STM (Scanning Tunneling Microscopy) instrumentation for manipulation at the atomic level.

Third, new technology waves are created from promises which require substantial resource commitments. There are gaps in time and place between the emergence of new technology waves and realisation of their full potential, so promises have to be made about what they will accomplish. Callon and Latour developed actor-network theory to explain how and why technological achievement is dependent on the alignment of various stakeholders' interests and strategies (Callon, 1986). The promise of new technologies may remain unfulfilled because, for example, of extreme difficulty in mastering it or as a consequence of strong social opposition. Gene splicing preceded the first recombinant drug by 10 years. The Internet was introduced in the 1960s, but it took 30 years for it to become widely used.[1] IBM first demonstrated nanotechnology in the 1980s, but public and policy recognition and significant support only emerged at the beginning of the twenty-first century. Actors may or may not wish to bet on the promises of a new technology wave and in so doing have to bear the risk of action or inaction for themselves. Their decisions impact on others' interests and fate. Financial markets provide the means and the tools for these bets and have indeed been most significant vehicles of recent new technology waves. Biotechnology as well as information and communications technologies have experienced the ups and downs of crowd mentality and speculation, building and then destroying jobs and vast amounts of wealth for investors. Nanotechnology may experience a similar fate and specialised venture capitalists were quick to position themselves in relation to it (Cheetham and Grubstein, 2003).

Fourth, hopes and fears arise with new technologies waves since questions, tools, context and tests that could assess the plausibility of possible predicted future do not exist and cannot be created without engaging in action. Each high technology wave generates intense expectations and debate in the spheres of science, industry and public opinion. Indeed, new technologies are stories in the making. Recurrent stories are built on how technologies will change the world. In relation to these stories about new technologies' potential, various types of actors (producers or users of these new technologies) are stakeholders who consider technology waves as risks or opportunities or both, depending on their location, role, perception, values and interests. All technologies involve both benefits and risks. For example, Al Quaida powerfully demonstrated how well-established technologies – knives and civil aircraft – could be transformed into instruments of death and destruction. New technologies are not immune.

As it is impossible to give precise definitive answers about the positive or negative impact of a new technology wave on society, resolution of such controversies may never be achieved (Latour, 2004). Individuals, social groups or organisations ask themselves the following questions time and time again: Will the new high technology wave be to my benefit? If it will happen anyway, how can I cope with it ? Such issues define the battleground for controversies. For example, a French patient's association (AFM) constructs an appealing story which relates a genetic defect to the gene-drug designed to cure it and recruits complementary actors (researchers, doctors, pop stars, TV, newspapers) to convey its story to the relevant target – potential donors (Kahane and Reitter, 2002). Meanwhile, competing stories are also proposed by other actors asking whether gene therapy is a good choice or whether money would be better spent on other approaches. Controversies can be found in relation to cloning, GMO and stem cells, and restrictions or the regulation of research is one of the outcomes; in relation to information and communications technologies, with issues such as privacy and the digital divide regularly discussed in the media; in relation to nanotechnologies, where matters such as the extent to which people can control nano-machines, privacy and toxicity, and these are increasingly debated.

In this context and to help them to decide on what to do, people and organisations make sense of the world on the basis of what they know, in the light of similar situations that existed and stories which they can use to frame their knowledge, questions and answers. Depending on their interests, beliefs and values, actors use stories and ideas to order the world and to convey and communicate to others their representation of the past, present and future (Weick and Browning, 1986). Myths based on culture, religion and history provide powerful stories since well-known shared stories are easy to interpret and communicate. Moreover, they have survived, and show the tendency to recur, and be recycled, probably because they address issues people face time and time again in relation to the uncertainties which arise in the context of new technology waves. We now describe a set of such stories.

Recurrent myths relating to new technology waves

Myths which are currently associated with technology-related risks are discussed below. They are embodied in mythical figures linked to Western civilisation, some with deep roots in Hellenistic and Judaeo-Christian culture.[2] The wide dissemination of these stories helps them to achieve instant recognition and reception in Western society. Interestingly, Western culture does not seem to have developed equivalent dramatic *figures* to support positive developments in relation to new technologies although it does produce celebrity technologists and heroes, for example Bill Gates.

Prometheus: Emerging technologies may destroy the existing order (of God(s) and/or Nature)

Prometheus is the symbol of a revolt against forces that are beyond human reach and of a new liberty gained once these are mastered. Prometheus stole the secret of fire from Zeus, brought it to humans and was punished for that (Aeschylus, 1990): Zeus arranged to have Prometheus chained to a rock where each day an eagle would come to eat his liver. To make this happen for ever, Prometheus' liver was restored each night to be eaten again the next morning. Prometheus' punishment was designed by Zeus to be everlasting so as to remind him constantly of his transgression. This myth tells people about the great divide that exists between their world and the god(s)'. In modern societies, the gods' order of the universe is still a powerful conception for some but 'Nature' has recently gained a similar status. The Prometheus myth tells us that transgression of the laws of the god(s) or Nature may have dire consequences which people may have to endure for ever, longing for their lost innocence. The story of Adam and Eve from the Christian Bible is based on a similar theme (Berlin, Brettler and Fishbane, 2003). In both Christian and some ancient pagan beliefs, acquiring knowledge is the characteristic of the entry into the human condition: it implies loss of innocence and the impossibility of returning to an idealised lost paradise.

In this perspective, developing GMOs, animal and human cloning, in vitro fertilisation, the use of embryo cells for therapeutic purposes may all represent such transgressions. The people who engage in technology development achieve new power but work outside the boundaries set by Nature, god and the gods: something for which they may be punished.

The greater the extent to which a new technology wave is perceived to transgress such boundaries, the more it is likely to be confronted by opposition and resistance. Resistance to and acceptance of technology correlates with the perceived extent to which the technology transgresses the natural order or god(s)' will.

Janus: Emerging technologies may carry a dark side as well as a bright one

Janus was the Roman God involved in changes from one state to another, or from one place to another. Janus is the symbol of transition, the defining moment or space at which something gives place to something else. Janus sits at the gateway between what is on one side and what is on the other: His character is that he is simultaneously both before and after the transition. Janus is represented as a god bearing two faces simultaneously and is like the two sides of a coin. Depending on how they look at him, people may see one or the other of these two faces although both are always present. New technology waves are linked to Janus since people cannot secure technology's benefits without experiencing its negative side. For example, increased connectivity of computers opens the door to viruses and databases might provide life-saving medical data and yet are also windows on our privacy. Mobile telephone technology involves antennae whose proximity has been claimed to bring health problems – mobile phones also invade our daily life and impose new norms of social behaviour in public spaces while also improving our connectivity. Genetic manipulation may provide new ways to diagnose and cure but it can also affect the diversity of humankind or help in the development of deadly viruses or bacteria for biological warfare. Nanotechnologies foster economic development and a new realm of applications for some but are also labelled 'necrotechnologies' by opponents (Le Hir and Cabret, 2005).

Since new technology waves offer benefits as well as risks, societies will face trade-offs, which will partly depend on the time lag between the perception of the technology's potential and its realisation. Society will favour new technologies which offer instant benefits and low risks while opposing those which may involve high risks and bring only long-term benefit.

Big Brother: Emerging technologies may deprive humans of their liberty and autonomy

Orwell gave readers of his novel '*1984*' a description of a world where all aspects of human life were controlled by a central dehumanised and unreachable authority, Big Brother (Orwell, 1950). As ICTs permeate our daily personal and professional life, the threat of Big Brother looms over how technology is received and perceived.

ICTs are present in many objects that communicate with each other and hold an increasing level of information on what we are, what we do and what we like. Already, credit card files provide easy ways to track users' preferences and behaviours. Whenever they are switched on, mobile phones can give information about where we are. Companies such as Microsoft, Google or Yahoo with their proprietary software are also often seen as potential Big Brothers because of the near monopoly status they have achieved. The drive to counter monopolies by open access development and non-proprietary technologies is also a clear indication of the reluctance of many users to be dependent on a provider with monopoly power. Nevertheless, users generally accept having life-related data tracked by credit card or mobile phone companies as a price worth paying for efficient services.

Depending upon the extent to which technology alters the balance towards or against the emergence of a monopoly, its reception will be smooth or will tend to generate intense controversy. The social reception of technology is negatively related to the concentration of power it may lead to and positively linked to the increase in capacity and ease of exchanges (speed, cost, amount, quality) it may generate.

Frankenstein and the Golem: Emerging technologies may spiral out of control

In both books *Frankenstein* (Shelley, 2004) and the *Golem* (Wiesniewski, 1996), people create near humans from inert materials to solve some of the problems they face but are not able to deal with. If Prometheus emphasises the separation and thus the transgression between the worlds of the gods and of people, Frankenstein and the Golem both deal with the separation that has to be maintained between living and non-living things. Both these components surround humans but mixing one with the other is a serious transgression that results in dire consequences, not because of god's wrath but because the man-made creature escapes the control of its master and turns against him. Living (as opposed to non-living) beings have the built-in capacity to evade human control and to act on their own behalf. Once the bottle is open, there is no way of putting the genie back into it again and great destruction may occur. Opponents of new technologies are particularly keen on such stories since they play on one of people's greatest fears. If modernity is about exercising control over our environment, our life and our destiny, then taking the risk of jeopardising it or having it destroyed by our own action is probably the worst nightmare humankind may experience. This preoccupation is particularly powerful when 'snowball effects' arise from technology's potential for self-propagation. In this case, potential disruption is hard to stop or control when it has begun.

In the 1970s, the opponents of nuclear power were the first to build on the argument of self-propagation in order to make their point. Powerful fictions were constructed showing what a world of radionucleides could hold for humans (Merle, 1983; Tarkovsky, 2002). Unfortunately for humankind, some of our worst fears have been realised and may be realised again. Chernobyl stands as an example of technological disaster with more threatened. Similar concerns were also raised at the dawn of biotechnology when scientists themselves considered imposing a moratorium on experiments involving recombinant living organisms (McClean, 1997). Opponents of GMOs have referred specifically to myths labelling GMOs as 'Frankenstein food' to emphasise that they consider them as radically different from other food (see also Chapter 12). Opponents of nanotechnologies work on a similar basis. In a recent novel, exploring the snowball effect linked to nanotechnologies, multiple nano-devices are combined with a self evolving software that enable them to self (re)assemble in order to perform complex tasks. (Crichton, 2003) One of the first achievements of these nano-devices is to organise their reproduction, the second is to confront humans that created them but who may threaten their 'life' potential. All these stories play on the potential of human-made things or creatures to escape and destroy the world of the masters who made them and to spiral beyond control. This issue is at the core of the legislator's precautionary principle which asks technology producers to provide proof that their technology is not liable to harm part of the society even before it has been developed or implemented. The incorporation of auto-assembly, self-propagation or auto-dissemination capacities fosters opposition to a new technology wave.

Dracula: A new technology wave and increased information and knowledge about technologies may kill human life

Dracula in his castle in Transylvania was known to be a dangerous seducer: a very attractive man whom young girls could not resist (Stoker, 2003). New technologies are often portrayed by opponents as seductive but dangerous: those who fall in love with them may suffer death and destruction. The Dracula myth is well illustrated by the asbestos crisis. After the Second World War, asbestos became the main material used for fire insulation. By the 1970s, most buildings and many ships incorporated this material which was thought to be safe. Indeed, it was found to be safe by classical toxicological tests until asbestos fibres became an issue. After years of exposure, professional workers manipulating asbestos fibres started to develop cancer lung pathology (mesothelium). This steadily led to recognition that asbestos was dangerous not only for workers but also for people working or living in buildings which incorporated asbestos in their construction. Some organisations tried to escape the consequences by denying any hazard from asbestos. This led to a situation in France where it is now recognised that 50,000 to 100,000 people are likely to die in the next 30 years from previous exposure to asbestos (Prieur, 2005). In addition to the damage to human health, building users and their insurers suffered substantial financial losses. A high proportion of the risks associated with asbestos had been passed on by insurance companies to reinsurance companies. Reinsurance companies now see nanotechnology as too powerful to resist. However, as a consequence of their disastrous experience in relation to asbestos and taking account of the Dracula argument of technology opponents, these companies have recently been giving serious thought to ways of preventing the repetition of a similar scenario in relation to nanomaterials (Swiss Re, 2004) while creating an opportunity for a new market and their activity.

The Dracula myth also relates to new technology waves in a second complementary way. Dracula could live only at night: light would kill him. Each morning, before dawn, he had to go back to his coffin in order to survive until the darkness of the next night. In their campaigns of resistance to new technologies, opponents often dismiss the tentative explanations provided by new technology promoters as biased and inconclusive. By making the technology's performance and potential an issue and by increasing controversy, they aim to stimulate fear of the technology.

The nature of technologies involved in a specific new technology wave determines which of the myths described above is likely to predominate. The next section attempts to provide some explanations for this.

Opponents and narratives in the context of new technology waves

Below we highlight three points that help explain the recurrent interaction between myths and new technology waves. First we deal with the emergence of opponents, second with the reduction of complexity in short messages and third with narrative theory.

New technology waves promote the development of opponents[3]

New technology controversies are often presented by technology producers as a battle between modernity and tradition. They see recurrent resistance to new technologies by organisations such as Greenpeace and left wing groups as merely a sideshow – a bitter but doomed reaction in a battle that modern society has already largely won. Most hard science researchers, industrialists and policy-makers in Western societies firmly believe that new technology is creating a more prosperous and free world than ever before. It is inconceivable to them that there could be a valid choice for society to live in a more traditional world devoid of the technological and social threats and benefits new technologies may bring. Opposing new technologies is seen as resisting the coming of spring. In the name of progress, those who promote new technology are determined to destroy the otherwise 'backward' way of life of those they see as discredited remnants of the past.

For example, many companies have a financial interest in installing personal computers and Internet access in every home and office. Nevertheless this objective has been accepted by governments and nearly everyone else in America, Europe and Asia as a progressive idea that would make societies everywhere richer and more democratic by increasing the availability of information and knowledge. However, the widespread dissemination of personal computers and the Internet destroyed millions of jobs in the printing industry, as well as the jobs of millions of clerical and secretarial workers. This was thought to be a price worth paying because these new technologies bring progress and create other types of jobs. But where is this better world and is it better for everybody? Where does progress lead?

These are questions of myths and values. Values are socially invented and maintained across generations, but they do not evolve as fast as technology. Since the Enlightenment and the modern scientific revolution unseated traditional and religious values as dominant intellectual forces in Western society, material and social progress has replaced values as the principal driver of the evolution of life and society.

Since societies are held together by tradition and the sometimes slow pace of social life, promoters of new technology are often fighting a war against tradition. Why should they be surprised when the defenders of current society strike back? There is a major misunderstanding here. Modern science and technologies are the product of science, technology, history and culture. Societies are what they are because of their past and current changes, and those who promote new technologies are at home in them. The modern world was created and belongs to society at large. But when interested parties try to impose new technologies together with ideas which contradict the fundamental values and assumptions held in other parts of society, it raises the question: Is this progress?

Complexity reduction in communication about new technology waves

New technologies are associated with uncertainties about their outcome. Science studies argue that the success of new technology often relates mainly to the battle for hearts and minds rather than to the technology itself. To succeed, new technologies have to be socially robust (Giddens, 1977). If this is the case, communication between potential stakeholders (whether supportive, neutral or opposed to the new technology) is crucial and will influence the outcome. But in controversies, complexity of new technologies and related issues tend to concentrate in short iconic messages.

In the traditional view, argument is the basic way people will make judgements because people often act like lawyers (Mason, 1969) waging legal battles in courtrooms. But if the concept of argument emphasises the transmission of information to receivers, the creation of meaning in perceivers may be of greater importance in the battle for hearts and minds (Delia, 1970). Indeed, when information circulates, it not only adds to or subtracts from an argument, but it is also changed qualitatively, made more or less meaningful in a process where a small amount of information may generate thousands of reverberating inferences constituting a pattern (Bowers and Bradac,1982). March and Simon (1958) note that uncertainty absorption is predominant in such situations. Uncertainty absorption happens when 'inferences are drawn from a body of evidence and the inferences, rather than the evidence itself, are then communicated'. In such a context, March and Simon point out that people who are not at the origin of communication can only make limited tests of these inferences: people may choose to refer to stories they know, so as to judge the validity and potential impact of the new technology for their lives and for those of others. Stories enable people to translate impressions of a distant event into a form that allows them to grasp its significance (Martin and Powers, 1983). This may explain the paradox in communication that the richer the information (e.g. in political elections), the shorter the messages will be without incurring any decline in effectiveness (Jamieson, 1985). Indeed, myths can deliver short messages very effectively without excluding rich information.

Myths as the ultimate metaphor for narration and 'nar-action'

If meaning and interpretation are key, Fisher's attempts to articulate a narrative paradigm may help to explain why mythical stories are powerful tools in the context of new technology waves (Fisher 1985a, 1985b, 1989). Fisher's narrative paradigm asserts that

> no form of discourse is privileged over others because its form is predominantly argumentative. No matter how strictly a case is argued, it will always be a story, an interpretation of some aspects of the world which is historically and culturally grounded and shaped by human personality.

This needs to be simplified – here is a suggestion:

The narrative paradigm has a certain logic (Fisher, 1985a), providing a synthesis of two strands in communication, the argumentative-persuasive and the literary-aesthetic strands, centring on the theme that reasoning is more than just implicative and inferential. In other words, stories are everything. People judge the reasoning in stories by how well the story hangs together (narrative probability) and how fully it rings true to experience (narrative fidelity). Importantly, Fisher suggests that in the rational paradigm that underpins argumentation, only experts can debate with experts, which means that non-experts are spectators. But this is not the case in the narrative paradigm where experts and non-experts are both story tellers and story listeners just like everyone else. They are both active participants in a battle of narratives (Ezrahi, 1998) where divergence, convergence and hybridisation between stories may occur.

However, new technologies are not only the results of discourse they are also products of action. Action matters as much as narration because it impacts on the imagination, symbols and reality of those humans who are or may be involved in the outcomes. This drove us to define, 'a nar-active paradigm' in the context of new technologies (Kahane and Reitter, 2002) in which both narration and action complement, test and reinforce each other in an interactive and iterative way. This brings about scientific and technological reality and context. In the nar-active paradigm, both narration and action have a symbolic dimension that human minds can read and interpret.

In this perspective, myths act as powerful resources for arguments both for and against new technology since they align both with the narrative paradigm and its reliance on symbols through discourse, and with the nar-active paradigm transformation through action. Moreover, nar-action opens the way to a dialogic perspective (Todorov, 1966). People who are exposed to myths identify with principal characters in the stories as if they were in their position and experiencing their dilemmas. In the context of new technologies, people not only try to imagine how they would themselves behave in the mythical context, but also try to work out what they and others would do and/or what could happen to them and to others as a consequence of the new technology's implementation.

Conclusions

New technologies act in waves that shape our era and our life. Although different in content, each technological wave shares some common patterns in terms of its impact both on its producers and users and the stories which arise about them. As a consequence of such commonalities, it is possible to generalise about them to some extent. Each high technology wave poses similar questions of society. Through their disruptive power, each wave confronts people with uncertainties about the future. Controversies are inherent in new technology waves, in particular because a significant gap (whether time lag, geographical or cultural distance) exists between their emergence and the realisation of their potential. Because at a given time demonstration and proof are not yet available or possible, humans have to turn to other mechanisms to create meaning and understanding for thinking, decision and action.

Opponents and supporters of new technology engage in a battle for hearts and minds, and to win resources, that will help to shape the future. Predictions of the impacts of new technology are inherently unreliable, built on promises that can be considered as mythical until their potentialities come true. Promoters of new technology present still unproven positive stories in order to gather support and engage in experimentation but they will often be challenged by alternative unproven stories, some of which show continuity with the past with recycled themes such as we have identified (Prometheus, Janus, Frankenstein, Dracula). These stories share a set of characteristics: they concentrate on deep questions humans face relating to their life and destiny; they achieve instant recognition and thus reception; they offer room for identification and interpretation; they are widely disseminated.

Myths are relevant and mobilised by actors in the context of new technology waves because of their historical legitimacy and their widespread recognition and dissemination. Innovation as an interactive nar-active process is influenced by what actors do or will do but as well by what they tell about themselves and about what they do or intend to do.

We suggest that technology waves are a characteristic of our era and that actors may benefit from understanding the myths mobilised in the context of previous technology waves. Future technology waves will share similar patterns with those of the past and present. Actors involved, whoever they are, will tend to use the same set of myths in relation to future technology waves and to their impact.

Notes

1. The World Wide Web Consortium was initially created in 1994 with the help of seed funding from the US Defense Advanced Research Project Agency (DARPA).
2. Similar or alternative myths shape non-Western culture but they are outside the scope of this chapter.
3. This section is a free reinterpretation of an article ('Traditional Culture Strikes Back') published on another topic (terrorism): see Pfaff (2005).

Part II Myths of Information and Communication Technologies

Chapters in this section consider the role of myth in creating and sustaining investment in Information and Communication Technologies (ICTs) and, at the same time, legitimating the directions in which capitalism exploits these technologies. While myths have been associated traditionally with classical legends and fables about the exploits of gods and heroes, Barthes has argued that myths may also be understood as the dominant ideologies of our time embodying sets of specific beliefs and structures of ideas and viewpoints. Thus the particular shape and direction of technological development defined by the capitalist economic structure and free-market ideology with which we are familiar is rendered obvious, normal and inevitable. The association of ICTs with images of revolutionary economic, social and cultural change is well documented in a diverse range of literature from government policy documents to the more populist accounts of futurology and cyberculture. The chapters in this section challenge some of the ubiquitous myths about the development and application of ICTs utilising a more critical framework which suggests that 'every myth serves an organisational and political function. Every myth seeks to control and shape some aspect of a social system' (Woodward, 1980: p. xvi).

Sharpe's chapter looks in detail at the fundamental engines driving the IT industry and the range of myths that are created, deployed and abandoned in efforts to support the demands for continued growth in the industry. He argues that there is not one myth but many which are deployed through the marketing and publishing media in order to keep markets primed for the next round of ICT products, and to influence the take-up of these products and services by policy-makers and the wider public. The impacts of this relationship are questioned, particularly in relation to the British National Health Service.

Stepulevage's chapter explores the implementation of large-scale, commercially developed, complex software systems known as enterprise resource planning (ERP) in the environment of higher education. Her study questions the myth that ERP can represent the reality of administrative work systems and that this software represents 'best practice'. The dominance of such myths influences the strategies and practices that underlie the implementation of ERP applications in higher education, and helps to explain the problematic experiences of those working with such applications.

Walker's chapter focuses on technological change in the media sector focusing on interactivity as the compelling buzz-word of the new digital society. It examines the discourse around new communication technologies and the debates they engender about issues of access, control, democratisation of the media and the empowerment of individuals. It challenges the technologically determinist myth of empowerment and democratisation which is a feature of some accounts of the potentials of new communication technologies and, by exploring both the historical development of technologies such as video and cable, and the new forms of digital interactive television, examines the economic and commercial imperatives which underpin the development and application of new technologies in the media sector.

Langstone's chapter addresses issues arising from the relentless growth of CCTV in Britain and the development of, what has been termed, a 'surveillance society'. With up to 4.2 million CCTV cameras in Britain, approximately one for every 14 people (BBC, 2006), this chapter examines the myths surrounding the introduction of CCTV technology and highlights some of the problems arising from assumptions made about its effectiveness in tackling crime.

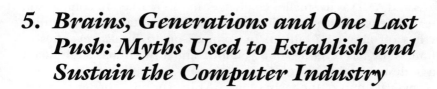

5. Brains, Generations and One Last Push: Myths Used to Establish and Sustain the Computer Industry

RICHARD SHARPE

Introduction

Information and Communication Technologies (ICTs) bear the stamp of the techno-economic system which produces them. This stamp skews their appeal and application towards the metropolitan centres of the globe, and towards certain types of users in the metropolis. Other users, on the margins of society in the metropolitan countries and outside the metropolitan centres, make what use they can, if they have access to ICTs at all, as a result of their own initiatives. Little is done by the suppliers of ICTs to widen their appeal beyond the demands of the richer countries and peoples.

First we look at the fundamental engines driving the industry. Then we look at how the central myths of the industry have been changed as the circumstances change. Finally we look briefly at how the myths of ICT are having an impact on health care in the United Kingdom.

Engines driving the ICT industry

Over the past several years, the rate of development of semiconductors in terms of performance and cost reduction has been truly phenomenal. This has been encapsulated in Moore's law which suggests that the performance of semiconductors doubles every 18–24 months. Networking has also developed very rapidly and its rate of development has been encapsulated in Metcalfe's law that the value of a network increases exponentially as the number of users increases, while the networking costs only increase linearly (Moschella, 1997: p. xii).

Both the hardware developments exploiting Moore's law and the networking investments exploiting Metcalfe's law demand ever-increasing volumes of capital. ICT developers are becoming increasingly capital-intensive. The demand for increasing volumes of capital has often led to a duopoly where a dominant vendor competes with a smaller ICT vendor or a set of them. The classic case is the role of Intel as the dominant vendor of micro-processors and AMD as its smaller challenger in this market. In comparison with Intel's write-off of over $20 billion a year in the early twenty-first century, AMD writes off about $6 billion. In comparison to Intel's revenues of $30 billion, AMD has revenues of $5 billion. Similarly, HP has a dominant position in desktop printers.

The physical limits of the technology are often not the barrier to development; lack of capital or the high price of capital is becoming the main brake on the speed of ICT developments. The demand for large amounts of capital, in its turn, forces the developers of ICTs to move as swiftly as they can into volume production. This is the only route through which they will recoup their increasing capital investments. By adopting this strategy, ICT developers are following the path laid down by Intel in the 1960s. Intel's strategy was, and is, to go for volume production as early as possible, to price the product for volume sales early on and to obsolete the product with the next one as soon as possible. As they ramp up production, the ICT developers also seek out any fresh markets which will adopt their technologies. Hence, the economic benefits of rapid technological change spread rapidly from their original base in specialist applications, to widespread computing, to networking, to mobile networking and to consumer markets.

Where ICT vendors do go head-to-head with a larger and more monopolistic player they find they are undercut. The larger player, spread over more markets, can use their dominant position in one market to lower prices in others.

De facto software standards can have an impact similar to the increasing demand for capital in the hardware industry. Thus, Intel's dominance is mirrored by Microsoft's. There are counter tendencies which make room for new entrants. Incumbent vendors focus on improving their own product line and can be caught out by radical changes in technology. Incumbent vendors can become hard to coordinate and add layers of management which are not carried by smaller enterprises. As creative destruction purges the industry of the less successful product vendors, a core of old and new is formed. For example, IBM is a remnant of the oldest part of the computer industry, pre-dating it with its Hollerith-based tabulating equipment of the 1910s. Burroughs, Sperry, Honeywell and ICL of the same computer era did not fight through successfully but are today amalgamated into other companies or have gone out of the industry.

This drive for volume production and ever-larger markets leads towards commoditisation: the production of standardised products with few discernible individual characteristics which are then sold mostly on price. Some vendors resist this as they see their profits disappearing: but all will cut prices to gain a larger market share when they can. Economic pressures force their hands. Those who refuse to play the game are shunted to the sidelines and fight a constant battle of survival, as we see from the histories of Sun Microsystems and Apple Computer, among others.

The PC market is a clear example of this drive to commodification. There are hundreds of different logo badges on PCs but the product is roughly standardised. Price and the channel through which the user can buy become the key attributes of selection. For example, the UK company Research Machines survives by concentrating on the education market. It cannot afford to go head-to-head with the leading PC vendors in the general market.

Forced to aim for largest markets

These economic drivers often force ICT developers to aim for the largest markets with the surest returns they can. These markets are in the metropolitan centres. The results from the largest ICT company in the world, IBM, clearly show the pattern. In 2006, IBM earned 39 per cent of its business from the United States, 11 per cent from Japan and 50 per cent from the whole of the rest of the world. This bears no relationship to the distribution of the world's population: United States 295.7m (4.6%); Japan 127.4 m (2%); RoW 5.9 billion (93.4%). Indeed, it is also more heavily focused on the United States than in the past in terms of GDP: United States $10.9 trillion (29.8%); Japan $4.3 trillion (11.8%); RoW $21.4 trillion (58.4%). IBM's results now match the pattern of world GDP more closely than in the 1980s when it won 60 per cent of its business from the United States. This shows the extent to which IBM has become a global operation rather than just a dominant US one.

Creating ICTs for users outside this mainstream of metropolitan markets and dominant sectors is risky. The ICT vendors will only go outside these dominant sectors in metropolitan markets when the profits from dominant sectors in metropolitan markets are too slim to survive. The profits are still good enough to sustain them, so these markets keep their attention. The public and the private sectors of the metropolitan centres have already reached the saturation point of one PC per desk. The business is now gained from upgrading existing users. And this cycle of upgrades is enforced by continuous technological development supported by continuous advertising and other forms of marketing in order to keep these markets profitable.

Therefore, ICT products are developed mostly with the infrastructure, cultures and applications of the metropolitan markets and dominant sectors in mind. A stable and reliable electrical power supply is taken for granted. Adequate climatic conditions for ICT products are equated with the climatic conditions of the metropolitan home and office because that is where these products are designed for.

ICT developers use technology changes either to reinforce their dominant role in these dominant sectors in these metropolitan markets or to try to destabilise the position of dominant vendors. These new technologies provide faster access – as in the current adoption of broadband. Or they provide higher quality than the previous technology – as with the adoption of the CD replacing the floppy disk and tape. Or they provide greater range – as does mobile telecommunications compared with fixed-line telecommunications. Or they provide richer content – as with the replacement of the CD by the DVD.

But these technologies do not diffuse into use in a steady stream. Early adopters may rush to have the latest, but until the adoption gets over a hump in early use and is taken up by a significant proportion of the potential population, they may for ever remain a minority technology.

No stable state

There is, therefore, no stable state of the market or of technology developments. There is no point at which all have equal access or equal technology. In addition, there is a restricted trickle down from dominant sectors in metropolitan markets to other markets. There is a delay of months between some forms of technology adoption in the United States, Europe and Japan. But this is not at all even, nor, one way: for example, the United States is not as advanced as Japan or Europe in mobile telecommunications or mobile use. Even the largest organisations in dominant sectors in metropolitan markets find that they are playing catch-up all of the time. They may wish that the rate of technology change would lessen, but as long as the capital is available and the physics of the technology allow, change will be continuous as they are faced with new challenges and opportunities by their ICT suppliers. And each individual business user would, in any case, like to have a new piece of technology which its competitors do not have in order to create advantages in terms of higher productivity, lower costs or a new product/service offering.

Yet central to the myth of the continued beneficial impact of ICTs is the notion of one more push: the problems of yesterday will be solved by the adoption of the technology of today. For example, it is only a matter of getting the price down far enough for every family in the consumer market to be able to afford a PC and thus to have the educational advantages which used to be available only to a few with access to computing and storage power. This was the cry of the 1980s and the early 1990s. Yet the PC is not enough. This steady state of relatively widespread PC use in metropolitan markets does not last for long. Then access to networks had to be added as the content shifted from delivery through storage media such as tapes and floppy disks to delivery over networks. That was in the later 1990s and into 2003. Then dial-up network access was not enough: the content had become so enriched as a result of the greater power available in servers which new generations of software could exploit that broadband access became essential. Everybody is, eventually, playing catch-up. Pioneering early adopters continue to search for the latest and trumpet their adoption of the very latest. Magazines and other media review the latest products, again a relative term continuously changing. Other ICT product and service vendors launch products or services which rely on the next range of technology, then the next, then the next.

There is no stable state which will level out the inequalities of the current times because replacement technologies generated by the techo-economic motors described above will create alternatives. As change is constant, driven by the constant charge to extract profits from the exploitation of Moore's and Metcalfe's laws, then a steady state is impossible: the only steady state is one of change. Those trying to reach a steady state are trying to hit a receding target; those advocating one more push delude themselves or others: there will be yet another, and another and another push beyond this one. The alternatives generated by the techno-economic motors will be more readily available to the already technically savvy, the early adopters, with the necessary money. This excludes those on the margins of society who are always at a disadvantage as a result. Not just once, but time after time, after time.

That, then, is the techno-economic context of the 'one last push' myth. Now let us look at the longer term: what myths have been used to open up the ground for ICTs? We will now turn to the development of the myths of ICTs and show how they have been changed to deal with changed circumstances. We will focus on the myths in the computing side of the ICTs.

The myth of the electronic brain

The first myth was that the computer would form the basis for the 'electronic brain', a convenient and successful image at the core of the myth which was used again and again in the later 1940s and early 1950s. This electronic brain would 'be able to solve problems in physics, electronics, the structure of the atom that no human being has any hope of tackling and, who knows, may one day give the answer to the question of man's origins'. This is according to one description of what the electronic brain could do (Comrie

1944 cited in Ifrah 2000). Comrie was describing the Harvard Mark I, an electro-mechanical programme-controlled calculator, partly funded by IBM and built in one of its plants. Success with the Mark I prompted IBM to cooperate with Columbia University to build the Selective Sequence Electronic Calculator. From the start of 1948 the SSEC was installed in the shop window of IBM's New York HQ. 'The [SSEC] calculator was taken out of service in 1952 after a short life, but long enough to fascinate the public, who came in their thousands to look at the lights on the calculator blinking away' (Ifrah, 2000: 314). Public demonstrations of the uses of these electronic brains firmly rooted the myth in the minds of the general public. Commenting on Comrie's use of the brain myth to describe the Mark I, Ifrah states, 'Thus we see how far an ill-chosen terminology has led us down the path of exaggeration when figures of speech are taken at face value' (Ifrah, 2000: 315). More importantly, aside from exaggeration, the term creates a digital sublime, an easily understood picture of what the future of this technology holds.

The myth of the electronic brain created the space in which these industries could emerge, catch the eye of investors and be justified for public funding in democracies. Many of the early investments made in the development of computers in this stage were made without a full cost-benefit analysis. It was impossible to calculate a cost-benefit analysis as the real benefits and the actual consequences, both the up and down sides, were unknown. After all, the early development of computers was funded by the state which would spend to ensure its own survival through invest-ments in code-breaking or command-and-control systems. And by the very nature of a new technology, the investments were fuelled by faith, the faith of the pioneers persuading the controllers of private and public funds that the benefits could be achieved. For example, only in retrospect do we know that the SAGE investment by the US government was successful. It might have failed, as far as understanding of the technology at the time could tell. The very strength of the myth of the electronic brain and the power of its image in the post–Second World War world of the victorious powers acted as a cloak for the failures and problems which inevitably occurred.

The myth of the electronic office

Generally, the development of computers was not a failure. Hardware became increasingly reliable, software increasingly specialist and complex. Proposers of computers spread out from their original limited technical base in code-breaking, scientific/technical calculations and command-and-control systems to the commercial world of data processing through pioneering projects such as Leo, the Lyons Electronic Office (Caminer,

Aris, Hermon and Land, 1996). The Leo computer, which ran the first office computer job in the world in 1951, takes the overarching myth of the electronic brain and applies it to the office. Now the office itself becomes electronic in the title of the system, although course, it is not. Instead the wages office, the first office to be affected by administrative computing, required less labour in order to make the calculations needed. But it did not disappear into electronics. The electronic office was a useful image rather than a reality. But myths are not, warns Mosco, to be judged by their approximation to reality, but by whether they live or not. They are also to be judged by those to whom they are useful (Mosco, 2005: 29).

The development of Leo is a clear example of developers creating the technology without clear cost-benefit analysis or rigorous technical evaluation. They cannot know that the system will work until it is built and tested. There is no way of knowing: Caminer, lead developer of Leo, clearly described how the developers of Leo struck out in the dark based on untested and untestable assumptions. 'An important part of the work of the Lyons computer team lay in unmapped territory, particularly in regard to input and output (I/O) ... It was assumed – although no classical Systems Office Research study was carried out – that magnetic wire or tape would be the I/O medium and, for that matter, the medium of secondary storage. The Leo team's assumption, untested in any way, was that electronic computing required magnetic services' (Caminer, 2003: 6). Even when there was some theory behind the design, the theory can be wrong. Caminer again: 'These vacuum tubes had speed advantages, but continually disrupted trials by their proneness to faults, including the appearance of charges on more than one cathode at a time, which was theoretically impossible' (2003: 7).

The eventual success of Leo in running the payroll for Lyons was widely publicised deliberately by Lyons. It allowed TV newsreel makers into the computer room. The head of development gave a talk on the BBC. More importantly for a British audience, Princess Elizabeth, as she then was, made a public visit to Leo as early as February 1951, when the designers ran a test program for her (Caminer, Aris, Hermon and Land, 1996: 36–37).

The original myth of the electronic brain was now being subdivided, reduced into a range of forms which could be applied to the increasing range of computer applications. In short, the myth had to be changed in line with the changes in the world. The myth was proving to be successful as it enticed hundreds of companies to order computers before they knew they would be economically useful. Writing of the mid-1950s, Haigh says of buying a computer: 'for many companies that was an easy choice to make. Hundreds of computer installations were ordered before the computer's economic value could be demonstrated' (2001: 78). The *Harvard Business Review* sounded a warning in 1955: the 'revolution ... appears to be off to a faltering start. Too much was promised too fast, with the result that many businessmen have grown sceptical of the entire electronic data processing field' (quoted in Haigh, 2001: 78). Eventually the success of a few installations became part of the myth which overcame this scepticism. The success of the myth in fuelling demand for computing led to longer term changes.

Specialisation

One of the major changes was the creation of computer specialists. This growing specialisation created a market for specialist publications which played an important role in the further development of the myth.

Specialist publications were created by publishers to sustain the myth, to create a community of interests around it and, primarily, to make money. The most successful was *Datamation*, a US-based publication widely read internationally and launched in 1955 as a monthly. Publishers saw the opportunities of this growing community. They often used the growing demand for computer specialists to found publications funded by the revenue from classified job advertising. In the UK, *Computer Weekly* was launched in 1966, the world's first weekly computer publication. These specialist publications were less likely to use the term electronic brain. But it remained in popular use, employed to tell a wider audience about the technology. For example we have the book *Computers: From sand to Electronic Brain* in 1961 through to *The Electronic Brain: How It Works* in 1969.

Investors began to reap a return from their investments in this growing industry. The successful generation of profits led to new efforts to make computer technology yet more reliable and yet more widespread.

The myth of generations

As companies entered into the market for computers following the pioneers they created the notion of generations: there was not to be just one wave of investment and one wave of technology but successive waves each based on a continuous development. By 1972, the only company with a long history of competing with IBM to make profits consistently from its foundation in 1961 was Digital Equipment Corporation (DeLamarter, 1986: 352). The seemingly continuous developments in hardware made computers smaller, faster and cheaper, spreading the areas of applications. The difference between IBM's and Digital's approach was that IBM's main computer lines were mainframes and Digital's were minicomputers. Digital had differentiated itself in the market for computers by cloaking itself in the clothes of the next generation of computing.

By the 1970s, even with the limited history of less than 25 years, the industry was talking about generations, looking into the past and making bold statements about generational change. It was on the verge of the fourth generation, according to the myth generators. Walker and Walker described three previous generations depending on methods of processing, software, hardware components and, importantly, the most significant controlling parameter (Friedman, 1989: 17). Walker and Walker's description of generations was a contribution to a definition of the fourth generation. There was little agreement on what was to be the fourth generation. One important statement from an IBM source said it would be 'a complex network of processors and data communication devices servicing many different types of user and use and inputting and extracting information from a variety of devices in a variety of data processing modes' (Infotech, 1971: 153). The combination of processors and data communications with various modes of input and output as described by IBM above for the fourth generation largely proved a success.

This created the space and momentum for the fifth generation. As with every change of generation, the proposers were taking a leap into the future without anything like full cost analysis. Developers cannot know that they will not hit some as yet undiscovered barrier to development. For example, heat generation could undo their efforts to make computers faster and smaller. They cannot know whether the necessary software can be developed in order to exploit the growing power of hardware. By its nature, development is speculative; it relies as much on faith, supported by myth, as it does on cost-benefit analysis.

This fifth generation would, its proposers claimed, exploit the limited success achieved by those who proposed the myth of 'artificial intelligence'. Hardware was now powerful enough, in the form of the minicomputer, to sustain an initiative to develop technologies which could simulate human intelligence characteristics of seeing, understanding text, and even of conducting abstract reasoning. The notion of the electronic brain, long abandoned, was revived and recast as Artificial Intelligence (AI) (Feigenbaum and McCorduck, 1984).

The Japanese industry and government, often acting in close harmony, were as eager to achieve success in the computer industry as they had been in the car and consumer electronics industries. It plunged into the development of the fifth generation in order to mount its challenge to the growing dominance of the United States in the computer industry (Feigenbaum and McCorduck, 1984: 99–130).

The myth of successive successful generations had fuelled aspirations, had created a digital sublime, which was unobtainable. Fifth generation projects consistently failed to achieve their objectives. Their failures and limitations were swiftly documented (Unger, 1987). Now the proposers of the fifth generation had created a crisis for the computer industry. Its steady development seemed to be coming to an abrupt halt. The notion of generations was shelved and not discussed again. What was of benefit in AI was now implemented in limited ways rather than as a generational shift.

But as the fifth generation projects failed, Moore's law was still attainable by the semiconductor industry. Engineers in the industry were still able to double the number of semiconductors in the same space of silicon every 18–24 months and so produce, in volume, denser memory chips and more powerful parts of computer processors. The breakthrough came with the creation of the first microprocessor in 1972 by Intel. The industry, especially in the United States, regrouped around this development and its potential application to a wide range of products from hand-held calculators to instrumentation and on into telecommunications.

The myth of the second computer revolution

A new myth was created to support these developments, the myth of the second computer revolution which would be based on 'new technologies'.

The promoters of ICTs achieved part of this change in mythology in the 1970s by establishing the notion of 'New Technologies' and their beneficial impact on productivity far beyond the limited applications of contemporary computer technology. A strong summation of this variant of the myth is in Evans (1979). Evans based his widely read book on a TV series he made for UK's ITV which broadened the issues out for a far wider public beyond the book-reading public and the trade press. Now the myth had reached the widespread and powerful media of TV. A fine selection of the contemporary myth-promotional material was made in 1980 by Forester. His selection shows how the first full-length feature article on this myth in the United States was in *Fortune* in November 1975. It was called *Here Comes the Second Computer Revolution* (Forester, 1980: 3). *Business Week* was well into the subject by 1976. In 1977, *Scientific American* was carrying articles by manufacturers, including Intel co-founder Robert Noyce, on how microelectronics were made. By March 1977, *Science* was devoting a special issue to the subject, taking up the notion of *revolution*. The introductory article was titled 'The Electronics Revolution' (Forester, 1980: 16).

The success of these myth promotions led to a long period of policy formation in which successive politicians and policy advisers of almost every type proposed policies for nations and regions in order to gain advantage from the newly established myth (see for example, Mackintosh, 1986) A particularly influential UK statement was by Barron and Curnow in 1979. Ian Barron was head of the government-backed microelectronics venture of the time – Inmos – and Ray Curnow was at the influential Science Policy

Research Unit at Sussex University. Industry and policy formation were merging rapidly.

These promotional efforts put the ICTs based on microelectronics on the agenda. Over the coming years, the terms to be used were subtly changed to adapt to the new waves of technological development: micro-chips, to new technologies, to informatics, to Information Technology in 1982, to ICTs in the academic world, to new media, and so on to dot coms.

The original myth of benefits flowing from the implementation of ICTs needs to be amended as these technologies are folded into the reality of socio-economic life and as they meet, fail to overcome or overcome technical and socio-economic limits. The original myths have to be shed in order to continue to open more spaces for the implementation of ICTs. The notion of the 'electronic brain' gives way to the notion of data processing; the notion of generations gives way to the notion of the second computer revolution. So the myth is subtly changed from its original form into a new form.

One last push

The new form of the myth we see used to promote ICTs in the early twenty-first century is the myth of 'one last push'. All that is needed for the current problems caused by ICTs, or not yet addressed by the application of ICTs, is one last push of technology: the next generation of technology, the new product, the latest version will solve the problems of the past. In this way each successive wave of technology, each new product and each latest version is hailed as a fix for current ills. Can't communicate between individual pools of productivity with PCs? The fix is dial-up Internet access. Can't download the rich content available on the web over the dial-up Internet access? The fix is broadband. Can't protect your data and system from intrusion from always-on broadband technology? Implement security software: the fix is the latest version of firewall, anti-virus, spam-detection and security software. Can't use the web effectively? Use a search engine. Can't get the results you want from search engines? Implement the semantic web.

So the cycle goes on linking one technology with another in a seemingly never-ending cycle of replacement. This is fuelled and supported by a wide range of publications whose sole purpose today is to review new products and guide the hand of the consumer or commercial purchaser. See *Computer Shopper* or any of the other publications which crowd the shelves of WH Smiths or Tesco.

There are three parts to this latest variant of ICT mythology. The first is that the new system will solve the problems of the past. This takes ICTs into areas not touched before. Examples from the public sector include medical records, a system for passports, and a system for electronic distribution of welfare benefits. In the private sector they include the application of systems to support and capture customer relationships.

Another part of the myth is the role of the next upgrade: the next version of the product will solve problems created in the implementation of the past versions.

Some technologies go almost in lock step, taking the strain. The clearest example is the relationship between the hardware base provided by Intel and the systems software base provided by Microsoft. Don't have enough power on your PC? Get a new laptop with a more powerful microprocessor and larger RAM. Not exploiting the hardware base you have bought? Get the latest operating system version. Running out of power because the latest operating system is bloated? Get a new hardware base. And so on. The names of the products may change: from 8088 in 1982 through to Pentium; from DOS in 1982 through to Windows XP. But the process remains the same.

This myth, that the next purchase you make, that the next system you implement or the next upgrade you take will solve problems becomes central to making sure corporate, government and consumer users of ICTs are willing to continually buy the products and services of the ICT promoters. The expectation of benefit drives the purchase.

The third part of the current myth is based on the seemingly obvious benefits of linking previously unconnected systems or technologies. This is the myth of integration, of convergence. Instead of the 'stove pipe' application of ICTs to separate applications, the whole should be integrated. Both data and transactions should be seamless in order to take advantage of the intellectual property the corporation or public sector body has already created. For example, the back-end ordering system should be integrated into the front-end web presence to create a seamless ordering system.

The second and third parts of the current variant of the myth – the advantages of new systems and the continued advantages of integration/convergence – create a new source of profits for the ICT industries. In addition to providing the products, the industry can now provide the services necessary to implement the new systems and achieve integration/convergence. This shields some companies partially from the continued ravaging affects of commodification of ICT products as they move into the 'higher value' services sector. For example, in 2006, IBM won 55 per cent of its total revenues from services, 25 per cent from hardware and 20 per cent from software. Twenty-two years ago in 1984, the proportions were services 11.5 per cent, hardware 72.5 per cent and software 7 per cent.

This drive for integration through the exploitation of Internet technologies has enticed the private and public sector into a new wave of investments. As in the past, developers cannot know that the current round of developments will be successful because they do not know that they will not hit some as yet unforeseen fundamental barriers, as happened, for example, to fifth generation computers.

We can see how these factors have had a direct impact when we look at the potential impact of ICT products and services on inequalities in public health care in the United Kingdom. This is a key investment in the heart of the expansion of public services in the United Kingdom in the early twenty-first century. The Blair government differentiated itself from the Thatcherite agenda, the previous dominant direction in British politics, by a focus on increased investment in public services. At the heart of public services is the NHS. Over £76.4 billion in 2005/2006 was allocated to the NHS.

The myth of the stable state to be achieved by one last push, the solution which the next generation of technologies or products will produce, has led government to adopt ICTs as a major path through which to deliver public services. The UK government has set targets for all government services to be online. Take the delivery of health services in the United Kingdom and the inequalities created and recreated as a result of the techno-economic system which produces ICTs becomes evident. Most use and rely on ICT products. A key health-care service indicator is the spread of patients who use inpatient services. In 2003, it started from 17.8 per cent of the population under 15; again 17.8 per cent for the 15–44 age group; rose slightly to 20.9 per cent for the 45–64 age group; and rose steeply to 48.2 per cent for those 65 and over. The take-up of electronic online channels was in 2003 almost exactly the reverse curve: 90 per cent of 16–24 year-olds had access and 3 per cent of those aged 65 and over. The demand for health care rises as Internet access falls in the population (Sharpe, 2004).

Over time this may change as the current younger Internet users grow older. But their access to whatever replaces the current services of the Internet is likely to fall off. Older people do not tend to upgrade their technologies as fast as younger users: many of the heavier users of health care when elderly will be living on reduced incomes from which there is little to invest in ICT products. And further, when we take a broader view, we realise that many of the elderly who would not or could not have access to the Internet today will happily use the telephone. But for their grand-parents, the telephone was the radical technology of its time available only to a few subscribers.

Age is not the only differentiator in relation to health services and ICT product use. 'The gap in health outcomes between those at the top and bottom ends of the social scale remains large and in some areas continues to widen... These inequalities mean poor health, reduced quality of life and early death for many people', says the 2003 DoH report *Tackling Health Inequalities*. The availability of health services such as advice over the Internet will actually reinforce this health inequality. Internet access was at

58 per cent of the population in 2003. The social groups ABC1 had 75 per cent access; C2s 55 per cent. Those receiving benefit 47 per cent; Council House tenants 46 per cent; the disabled 27 per cent. The highest groups with access were younger people in full-time work. Even many in jobs have no Internet access. These include people employed in cleaning, catering, driving, manufacturing and construction. And those unemployed do not have easy Internet access unless they have bought it themselves. Of those out of work, the unemployment rate of ethnic minorities can be twice the average rate for the white citizens, according to Labour Force Surveys.

In short, the government is using a tool, ICT, whose use is already heavily skewed towards the more prosperous, to tackle a job of health service delivery, which is more skewed to the less prosperous. The only way in which this strategic weakness can be overcome, if ever addressed, is by employing the myth of one last push. The information poor, the digitally disenfranchised, the Internet disadvantaged will be able to be lifted to the level of the information rich, the digitally enfranchised and the Internet advantaged by the efforts of one last push.

All this, then stacks up against the myth of rising access to the latest technologies through which increasingly goods and services will be available.

Conclusions

There is no one myth of the ICT technologies. There are many. They evolve: they are created, deployed and abandoned as needed to support the demands for continued growth of the industry in order to support the exploitation of Moore's and Metcalfe's laws. They are created to make space for new entrants; they are deployed through marketing and publishing media in order to keep markets in the dominant sectors and the metropolitan countries primed for the next round of products and services of the profit-making ICT industries. One myth replaces another as it fulfils its purpose. New terms are coined to make the myth relevant in contemporary circumstances. New images are employed to create a resonance among investors, policy-makers and the wider public.

6. ERP in Higher Education: The Reinforcement of Myths

LINDA STEPULEVAGE

Introduction

This chapter explores myths relating to the development of computer-based systems. It concentrates on the implementation of large-scale, complex software systems in a domain of ICT systems known as enterprise resource planning (ERP) and considers a number of case studies published between 1998 and 2006 in order to explore the myths and realities of ERP as a technology of choice in higher education.

The chapter defines and situates ERP, reviews rationales for its take-up and identifies key issues highlighted in implementation of these systems within the private sector. It then focuses on the implementation of ERP for applications in higher education in the United Kingdom and United States. The study focuses on myths that help constitute a belief that ERP can represent the reality of administrative work systems and that this software represents 'best practice' in these work systems. These myths are underpinned by a belief in the systematisation of knowledge about and the standardisation of work systems. They are reinforced in strategies and practices that underlie the implementation of ERP applications in higher education, while at the same time the myths help to explain the problematic experiences of those working with the applications.

ERP as the software of choice

Software used in work systems can be conceptualised in terms of two major groupings, those oriented towards management of the computer technology itself (systems software) and those oriented towards the carrying out of functions relevant to a user group, sector or organisation (application software). Application software may be in the form of off-the-shelf application packages, for example word processing and spreadsheets, bespoke software which is custom developed for a specific user group, and large-scale application software such as enterprise resource planning (ERP). Both application packages and ERP systems claim to be customisable, allowing the users of the software to tailor it to meet specific organisational needs. While the creation of templates for word-processing documents or selection of format for the graphic presentation of data are common ways to customise application packages, the customisation of ERP software is a more complex undertaking.

ERP software provides the wide integration of data so that many functions, for example accounting, financial control and resource planning share a database. The systems integrate data across an organisation and embed a standard set of procedures for its input, use and dissemination in real time across functional boundaries and enable end-to-end monitoring and tracking of transactions (Grant, Hall, Wailes and Wright, 2006; Kallinikos, 2004; Mabert, Soni and Venkataramanan, 2001). Mabert, Soni and Venkataramanan identify three key properties of ERP software. It is modular in structure, whereby the software can be used in various combinations so that organisations can implement all or a subset of the modules. This modular architecture in general coincides with what is considered to be the conventional functional segmentation of organisations (Kumar and Van Hillegersberg, 2000), but as Soh, Kien and Tay-Yap (2000) note, these embedded models reflect a bias towards Western business practices. Another property is that it is multifunctional in scope and allows the tracking of a range of activities such as financial results, sales and human resources. A third is that ERP is integrated in nature, meaning changes in data cascade throughout the various functional components of the system. The reasons usually stated for adopting ERP systems reflect these claimed properties. The most dominant reason identified by Mabert, Soni and Venkataramanan was 'to simplify and standardize IT systems' (2001: 70). It is believed that this sort of package helps organisations overcome the fragmentation of their various domain-limited information systems with a relatively unified, organisational-wide software platform (Kallinikos, 2004: 9). The second most common reason given for adoption was to have access to accurate information so that interactions and communications with external as well as internal stakeholders is improved. This 'accurate information' depends upon the integration of data across organisational operations in ways that reflect their functional interdependencies in real time (Kallinikos, 2004: 8). ERP deals with what are considered to be strategic priorities, and therefore is considered a business rather than an IT solution (Mabert, Soni and Venkataramanan, 2001). ERP systems are expensive as they standardly encompass not only software, but also costs for consultation, hardware and infrastructure, an implementation team, and training, and most of these large-scale packages require extensive post-purchase tailoring (Sawyer, 2000).

ERP software represents a fast growing, popular segment in packaged software sales (Robey, Ross and Boudreau, 2002; Sawyer, 2000). The various ERP promotional materials claim that 'these systems can link the entire organization together seamlessly, improve productivity, and provide instantaneous information' (Mabert, Soni and Venkataramanan, 2001: 71). Much of the business information systems literature published about ERP accepts it as the inevitable next solution and rather than questioning ERP as the system of choice, this literature can be characterised by its concern with guidelines for successful implementation. There is, however, a growing body of work that critically considers ERP in its wider social context and raises issues related to possible transformations in the work of an organisation (e.g. Cornford, 2000; Grant, Hall, Wailer and Wright, 2006; Kallinikos, 2004). For example, Kallinikos raises questions about the relationship between procedures embedded in the software and the constitution of human agency after ERP has been implemented (2004: 12–13) whilst Grant, Hall, Wailer and Wright explore the technologically determinist discourse of ERP with relation to its social construction in implementation.

What underlies the taken-for-granted assumption that large-scale packages are essential for organisations wishing to update their technical systems? This chapter argues that some of the myths rooted in the development of computer-based systems generally underpin the belief in ERP as the most suitable ICT system for large and medium sized organisations. One of the definitions of myth cited by Hirschheim and Newman (1991: 34) is 'an unquestioned belief about the practical benefits of certain techniques and behaviours that is not supported by demonstrated facts'. The next section discusses a number of beliefs rooted in the development of computer-based systems that critical readings of the development of these systems show cannot be supported in practice. The myths chosen are those especially relevant to the ERP imperative.

The mythical development of computer-based systems

In an early examination of computing developments, Kling and Iacono (1984) conclude their analysis with the following question: Are ideologies essential to give meaning to an organisational strategy of computerisation and to mobilise support? The ideologies they refer to are those grounded in a shared language and beliefs about the meaning of computing within the organisation (1223), and myths help to constitute these meanings. The dominant ideology identified in the case discussed by Kling and Iacono was a belief in the efficiency benefits of the new computing system to the individual, department and organisation even though no one could indicate the costs of operating the new system or the costs saved by using it (1224). This ideology of computer technology providing the best solution rests upon an underlying belief in the accuracy and efficiency of rule-based procedures embedded within computer software. Wagner, Scott and Galliers (2006: 267) argue 'that enterprise systems in particular embed within them particular ideologies that discipline *what work is* and *why it should be done*'. Before discussing the literature on ERP in higher education, I set the context for my analysis of ERP by reviewing some myths of computing technology and systems development that are relevant to the take-up of these software systems. These myths concern the modelling of reality and the nature of software development. I argue that the factors identified for achieving success draw on and reinscribe the myths discussed here.

Reality as a well-ordered system

The management-oriented literature and textbooks for students of systems development and, more specifically, database development, assume that the construction of new computer-based systems require developers to perform a rational set of activities, organised as a series of stages.[1] The key activities are requirements definition, analysis, design, implementation and evaluation. For packaged software the stages are modified to replace analysis and design with selection and purchase. These activities are rooted in a mythical model of systems development first conceived in the 1960s and called the systems development life cycle (SDLC). With the development of database management systems, data modelling became a vital tool in the analysis and design phases of the life cycle. Conventional approaches to systems development that depends upon this systems life cycle framework rest on the assumption that reality is an ordered stable system. These approaches aim to find a true representation of the world and then focus on building this representation rather than dealing with the world itself (Dahlbom and Mathiassen, 1993: 51). In database development, this representation is referred to as the conceptual model, and as Beynon-Davies notes, the preferred reading of databases is that they are models of reality. 'The assumption is that there is some objective reality, elements of which can be captured and represented (hence the tendency to use terms such as requirements *capture*)' (2002: 99). These elements are reduced to a form that is usable, a data abstraction that can be represented within database software. In ERP systems, this model of reality is said to represent 'best practices' and therefore 'best practice' can be adopted in the purchaser organisation to help them maintain a level playing field with their competitors (Gratton and Ghoshal, 2005 cited in Wagner, Scott and Galliers, 2006).

Historically an organisation's reality was conceptualised as a set of routine, unchanging operations such as billing and payments. Early systems analysis texts assumed one step of the systems life cycle called 'analysis' would deliver what was assumed to be a true and complete specification of requirements to replace what were conceptualised as routine operations. The specification would then serve as the basis for the design of this system. With the introduction of computer-based office systems in the 1970s and greater quantities of information made available to senior management, new systems expanded from routine tasks to tools for decision-making. The requirements for these tools were conceptualised as rational decision-making processes, where there would be less reliance on human intuition, judgement and politics, and more emphasis placed on the analysis of empirical data and the calculation of a 'best' solution (Hirschheim, 1986: 185).

Database management software, also developed in the 1970s, further broadened the domain of reality in that developers assume that in modelling the reality of an entire organisation database, systems developers can enable better access to data, more accurate data and sharing and integration of data sets. As with earlier computer-based systems, the database is intended to act as a tool for management to improve organisational productivity and efficiency (Beynon-Davies, 2002).

There is a tendency to treat this modelling as an unproblematic process, carried out by employing a strategy of functional simplification to reduce complex human activity systems to models of reality (Beynon-Davies, 2002). Ciborra highlights the focus on abstraction of reality and rationality in arguing that what both managers and practitioners have carefully left out of the current approaches to systems design and management is human existence (1999: 85). The development of computer-based systems is rooted in an ideology that privileges technical capabilities and neutralises or renders invisible the material conditions, location and social environment within which the technologies are embedded (Sassen, 2002: 366). Furthermore the belief that human thought is mere calculation underpins many applications of IT and enables the pretence that reality can be modelled (Angell and Ilharo, 2004: 46).

The business information systems literature not only presents ERP as offering a desirable model of reality, but one that claims to embody current best business practices. As Galliers notes, 'The myth is thus created that the adoption of an ERP system will enable an organization to transfer to itself the "best practice" industry knowledge of how best to organize various processes' (2004: 251). While the software claims to represent a 'best practice' reality, business-oriented research on the realities of ERP outcomes is mixed, with opinions stating that it can be an asset that delivers on stated promises or a liability incurring high costs (Mabert, Soni and Venkataramanan, 2001: 60).

ERPs and reality

The assumption of an objective reality that can be modelled and the modelling of decision-making as a rational process, what Dahlbom and Mathiassen refer to as a 'formalization imperative', have a strong influence in the construction of complex software applications (1993: 21). The tendency towards rationality and formalisation is evident in the literature that identifies ERP's claimed benefits, such as 'the embedding of tacit organisational knowledge explicitly, in well-documented information structures and decision rules' (Davenport, 2000). Case study research also identifies this tendency, for example where a variety of locally negotiated grants management practices were rejected, and a corporate financial planning tool was inscribed as ERP-based best practice (Wagner, Scott and Galliers, 2006: 268).

As various case studies show, there may be a mismatch between the organisation's working reality and the software's set of rationalised procedures. These mismatches are evident to those working with the software, but they are not always evident to the purchasers of the system and to management. End-users make do by using shadow systems or other types of work-arounds so that their bosses receive the outputs they require. In the daily work activities of many users groups, mismatches are an everyday reality that they must deal with, and their ways of dealing with them relate to another myth, that of a design-use opposition.

Software development as a discrete activity

The life cycle model moves from problems to solutions and focuses on ways to achieve a successful outcome. Its conception of technology production is based on the myth that there are creators of new technologies and recipients of these technologies. 'The fact that this myth belies the lived reality of systems development and use has so far gone largely unchallenged, as has the simple designer/user opposition that underwrites the myth' (Suchman, 2002: 92). Key to this myth is that technical expertise is considered the necessary and sufficient form of knowledge required for the production of new technologies (93), while on the use side of this boundary, training in the operation of a new computer-based system is considered to provide sufficient knowledge for its effective operation (Stepulevage, 2003).

User involvement is considered beneficial to information systems development, but this usually manifests itself as the designer extracting facts from the users, allowing users to experiment with mock-up systems (prototyping) or having a user lead the project team where their involvement is continual, but interpreted very narrowly (Hirschheim and Newman 1991: 34).

This 'user' may be a worker who interacts with the computer system in the case of prototyping, or a manager who leads a project team.

Project teams developing software on the design side of this boundary share a vision of the software as product(s) with the user at a distance. Success measures for this software relate to profit, market and mind share rather than user satisfaction and acceptance (Sawyer, 2000: 50). Suchman terms this world of technology production as 'design from nowhere'. It is a world in which anonymous and unlocatable designers deliver technological solutions to equally decontextualised and consequently unlocatable users. She identifies this stance as closely tied to the goal of construing systems as commodities that can be stabilised and cut loose from sites of production for large-scale export to sites of use (2002: 95).

ERP, design and use

The simple dichotomy of design and use masks the complex social relations that cross these constructed boundaries of 'designer' and 'user'. The boundary between design and use has a self-evident material presence with ERP software since these systems are developed offsite, and all the activities that take place in the purchasing organisation are termed 'implementation', the stage of the systems life cycle in which the software is tested and users receive a functioning system from developers. Misfits, the gaps between package functionality and organisational requirements, are a common problem with packaged software, however. In the case of ERP, there are mismatches at the detailed process level that create considerable implementation and adaptation problems and require tailoring onsite (Kumar and van Hillegersberg, 2000: 25; Sawyer, 2000).

These tailoring practices blur the distinction between designer and user in that identity of one of the user groups in the purchasing organisation may be an IT professional. The internal IT professionals may have had little or no say in the purchase of the software and may or may not have been involved in the detailed analysis of requirements, but they may be required to help configure the systems to the organisation's work practices even though they may not have the detailed knowledge of the software (or work practices) to do so. Nonetheless, examples from case study material show that interactions between internal IT staff and end-user groups reconstitute this design-use boundary so that user groups continue to be constructed as receivers and IT staff as designers of these ready-made systems. Case study examples, however, demonstrate that user groups blur this boundary in their attempts to fit these technical artefacts into their local work context. They do so by changing menu screens or requesting additional functions. Some of these embedding activities, referred to as local configuration, customisation or maintenance are often conceptualised as discrete phases in a systems life cycle, but more critical studies argue that this work of fitting or embedding the technology is another form of design work (Clement, 1991; McLaughlin, Rosen, Skinner and Webster, 1999) and that user groups must continually act to appropriate new technical artefacts into their sets of work practices and their current environment (Suchman, 2002: 93).

Context for implementation of ERP in UK higher education

The external environment of HE has been rapidly changing in the last decade. The number of students studying at HEI has increased, and there are declining per capita resources. There is also an increasing demand from the state for accountability, for example reporting of statistics and audits. Internationally, universities are competing for students and research funds, while locally and regionally, universities have been encouraged by the state to support the local community in technology transfer and transition to a 'knowledge-based' economy (Cornford, 2000: 511–512).

HEI's response to these pressures has been the growth of a new managerialism in which 'technology initiatives appear to be generating pressures for the establishment of a more "corporate" form of organisation where goals, roles, identities, abstract rules and standard operating procedures are made explicit and formalized' (Cornford, 2000: 515). Organising principles are moving from those of collegiality, characterised by consensus building and a democratic decision-making process within committee structures to one of direction from the centre, characterised by an information strategy based on a mechanistic rational approach taken from the private sector (Allen and Wilson, 1996). This push for formalisation can be seen as setting the framework for ERP with its promises of cross-functional information integration and free internal information flows (Allen and Kern, 2001).

The 'information strategies' of HEIs tend to focus on the technology and technological systems in isolation (Allen and Wilson, 1996: 247). The software is conceptualised as a complete system and management decisions about levels of customisation are seen as local alignment of software and processes. In the following sections, case studies of ERP implementations in HEI are used to explore the myths and realities of ERP implementation.

Representing desired realities

A research study of justifications for ERP adoption by universities in the United States and Australia shows that improved processes were an important rationale, whether HEIs had an awareness of potential efficiency improvements promised by an ERP or not. There was an underlying assumption that ERP systems could give HEIs what they desired, and managers were willing to change their processes to the supposed 'best practices' of ERP. That ERP systems embody the myth of 'best practice' in fact offers the assurance that the university's decision-makers are acting rationally in choosing the best option (Oliver and Romm, 2002: 207).

As noted already, the conception of reality in computing applications such as ERP is grounded in the idea that work can be and is done by following sets of rules. A case study of an ERP project in a UK university demonstrates an attempt by software consultants to impose these rules. Consultants presented a set of 'workflow process diagrams' to show how step-by-step sequences of events, such as setting up a research account, should be carried out to integrate with the procedures programmed in the software. Parts of the process were designated as taking place 'on the system' and parts 'off the system' with constraints on who undertakes which tasks and in what order (Cornford, 2000: 517). What became clear as the diagrams were discussed was that 'there is more than one way, in current practice, in which a particular step in the process can be handled' (518). When there was no local resolution of differences between various people's way of setting up accounts and the way imposed by the software, a policy decision on how to proceed was passed on to the university management. Cornford concludes that this sort of resolution not only sees the tightening up of roles and procedures locally, but across the university as a whole, constructing a more corporate institution to host the software. As a director of information services said in interview, 'The thing that trips [the implementation] up isn't that the technology doesn't work, it's trying to recreate the organization so that it can usefully apply the technology, rather than crippling [the technology] so that we can do things the way we did before' (Cornford, 2000: 519).

This asymmetrical alignment between the software and processes can be conceptualised as the reconstruction of reality. It is a reality strongly influenced by the representation of reality embedded in ERP and also evident in two other case studies dealing with ERP implementations, one in the United States and another in the United Kingdom.

The negotiations involved in an emerging ERP system in a US elite ('ivy league') university demonstrates the strength of the ERP vision of a new world order. This developing ERP system was bound up in a partnership between university management, led by a corporate minded VP, and one of the big software companies. It was hoped that this partnership would produce 'the definitive global standard, a gold standard, for academic administration and thereby influence future development of universities around the world effectively colonizing the unknown global future' (Scott and Wagner, 2003: 299). In other words, this university intended to produce the definitive HEI software that would constitute the new 'best practice'. The software in practice, however, demonstrated that the partnership's reality-product of global best practice could raise the question 'best practice for whom?' Staff groups found that they could not retrieve information fundamental to their ongoing activities. One example given was that of a research user group. They found that the simple yet essential question 'how much money is left to spend' was not available from the reports generated by the software after the first phase of the implementation (304–305). This situation could be interpreted as a case of the local reality and the reality of the corporate software being at odds. A member of the local core group of managers is quoted as saying,

> The difference between now and back in the days of home development is *we're not our own masters*. Our code is provided by vendors who have their own agendas. [Ivy] cannot dictate how it wants to do its business *by* itself. ... the world keeps changing and there's nothing you can do about it – gotta keep up – change with it. (302)

Another case study, based in the United Kingdom, explored the possibilities for end-user participation in the embedding of software for supporting the administration of student records at an 'old' university. It showed the dominance of the ERP software in a reconstruction of reality for department administrators (Stepulevage and Mukasa, 2005). The management information systems (MIS) project team seemed to position the software as the central and authoritative actor in the implementation and the software construction of the activity was privileged over the administrator's knowledgeable assessment of registry needs, as this administrator's comment shows, 'We produced reports and some of them were all wrong... [MIS] don't understand registry work and when you ask them...they say, why don't you change the way you do things? (195).

The administrators' reality, however, was grounded in a larger social system, not a comprehensive and well-ordered set of procedures that integrate data collection and processing. In the administrators' experience, the ERP took little account of how the work is actually carried out. An important insight into the gap between the imposed reality of the ERP and the dynamics of local work activities was made by one of the administrators in discussing problems with the system:

> Fortunately I go beyond the standard reports. ... Standard reports are limited in what they can do, for example I need to find out how many students in the departments are taking our modules so to get that information, I have to use [RG]. Also not everybody wants standard reports. (Stepulevage and Mukasa, 2005: 117)

The examples drawn from these three case studies show that there were mismatches between the work systems as envisioned by management, the procedures embedded in the software, and the work practices as conceptualised by user groups. The next section examines one of the case studies to raise the question of whether the mythical opposition between design and use can be transcended through the appropriation of ERP technologies.

Implementation as ongoing design

There is evidence to show that management are able to declare ERP adoption a success even though there are many, and sometimes insurmountable, problems in implementation. Grant, Hall, Wailes and Wright (2006) offer a number of examples in their study of Australian organisations' adoption of ERP where the software is deemed a success even though end-users needed to develop work-arounds and for one function, had to disable a module in order to get their work done (7–9). In another case study of an Australian university, management was considering the purchase of an ERP system, but in negotiation with a vendor they recognised the extent of the mismatches with their 'idiosyncratic' needs and decided to instead opt for a bespoke system (11–12). This section draws on data from the case studies to show that these mismatches can sometimes provide the occasion for transcendence of a boundary between design and use. That ERP would and/or could not be accepted as a new reality was evident in detailed examples from the case studies. In both the US and UK universities, the inadequacies and limitations of the software were overcome by user groups who had a commitment to getting the work done. In both cases, however, the myth of a new reality served its purposes for those who promoted and/or purchased the software.

Customisation of the software is one of the features of ERP that enable the software to be configured for a variety of organisational settings. In the UK university, the formal decision-making regarding the customisation of the software was dealt with in a top-down manner, with management requesting modifications. Administrators suggested, asked for and/or demanded changes to software as well, but neither external nor internal implementers negotiated with them to develop alternatives. Instead they attempted to impose the new reality of the ERP. When the manager was interviewed, he stated that he was satisfied with the software, but all the administrators interviewed expressed dissatisfaction with what they experienced as a very inadequate system. The myth of software as constituting a successful new reality for management co-existed with the profound dissatisfaction of some of the end-user groups, with half the admissions team leaving their jobs (Stepulevage and Mukasa, 2005: 196).

The focus of the case study based in the UK university was on the lower-level administrative staff and their opportunities to participate in the development of the new work system. When asked about their involvement in the implementation project, these workers expressed a sense of marginalisation, with one administrator saying they are not even informed about enhancements. She gave an example of coming in one morning and 'all of a sudden screens were changed' (Stepulevage and Mukasa, 2005: 193). Administrators attempted to negotiate with MIS staff and when these negotiations were unsuccessful, that is MIS staff refused to or were incapable of making changes, administrators developed work-arounds. One of the administrators who had developed a set of reports from data exported from the ERP and into a spreadsheet application made it clear that embedding new software applications into her local work practices was a standard activity as far as she was concerned, saying 'I am very used to trying to use computers and trying to satisfy the needs of my boss who wants specific information' (196). While administrators demonstrated that they had carried out design work in order to embed the software into their work systems, their narratives also show that a design-use boundary was continually reconstituted. This boundary was moveable and dynamic, with IT professionals within the university being in a pivotal position. They were a user group in relation to external consultants, but acted as designers when relating to local management or administrative end-user groups.

In the US university, the authors identify what are termed 'trials of strength', key points at which different actor networks negotiate and some gained greater influence in shaping the emerging ERP system (Scott and Wagner, 2003: 291). In order for the ERP to become successful, the authors concluded that it was 'a present day "matter of fact", a part of ongoing work experience, and the work involved in making it function has to be reluctantly absorbed by those whose work lives it touches' (287). This reluctant absorption is evident in the interactions between user groups, the project team and local management. The positioning of the project team in this case study demonstrates the 'design from nowhere' referred to by Suchman.

In developing a product that represents a global standard for academic administration, the external software company was considered to have understaffed the project, with software experts rarely working onsite and programmers unreceptive to learning the university's business (302). This is evident in one of the trials of strength identified by the authors. It concerned report development within the medical school faculty. The VP and ERP developers envisioned a new more corporate way of working and programmers embedded a set of procedures based on time-phased business planning for dealing for research administration. From the project team's perspective, the faculty's way of doing things was rooted in an outdated checkbook mentality that was in conflict with 'the vision' for the ERP system (305). As the first phase of the design was closing, however, the faculty demanded that existing legacy reports remained live until new tools were developed to give them the totals they required (305). The project team developed a bolt-on to the ERP software, which the research administrators considered inefficient and cumbersome. With development activity that almost parallels that of the administrators in the UK university, a shadow system emerged: 'Resourceful administrators exported data from the bolt-on and imported it into [a spreadsheet application] enabling them to recreate silos of activity', that is tables of data about funding availability, and so on (306). The newly constructed reality of heterogeneous shadow systems spread around the university, and three years later in a follow-up e-mail interview, a member of the research faculty user group noted 'Starting over was not an option (apparently) so we still do not have either system fully in place. No [check-book] replacement, no fully developed [ERP] approach' (306). As in the Australian university referred to earlier, this group has decided to design a full-scale alternative to the ERP grant reporting system for themselves.

Conclusions

Kling and Iacono's questioning of the necessity of ideologies is relevant to the case studies discussed in this chapter. ERP software has become a dominant organisational strategy of computerisation. Why ERP is the preferred option cannot be explained by reasons such as 'to simplify and standardise IT systems'. As the case studies show, simplification and standardisation may be intended, but it is not always realised in practice. The concept of myth, defined earlier as an unquestioned belief about the practical benefits of certain techniques and behaviours that is not supported by demonstrated facts, offers a richer understanding of why ERP software is the preferred technological choice (Hirschheim and Newman, 1991). The benefits promised by ERP are optimisation of resources, competitive advantage, business efficiency, best practice and most importantly a shortcut to the accomplishment of organisational goals. The Ivy University case study clearly demonstrates decision-making based on a belief in the benefits espoused by ERP developers even though they were not supported by demonstrated facts. A decision was made to embed time-phased business planning even though a working alternative was in place and was favoured by those using it, and despite any evidence that the new method was tried-and-tested within university administrations (Wagner, Scott and Galliers, 2006: 264). The design and use-opposition in this case exposes the dynamics of power and authority in decision-making. The organisation context and power relations between the different actors helped reconstitute the myths of ERP in that a work-around was implemented in an effort to satisfy local user groups, while the vendor developed 'best practice' ERP software for the global HEI market. The other case studies also demonstrated that efforts to overcome the deficiencies of the system actually reinforce the myth of its success. The mismatches that emerge when ERP systems are implemented are usually dealt with by workers who had no involvement in configuration decisions and did not have an unquestioned belief in the benefits of the software.

The decision-makers who invest in ERP reinforce the myth on a larger scale as they aim to be seen to take on industry or sector standards or to meet funding regulations. In the case of Ivy University, they wanted to work with vendors to set a definitive global standard for academic administration. When this 'global standard' ERP system was placed in the market, senior management needed to act as delegates for the vendor's system in order to protect their interests, as they had made a substantial investment and feared the possibility that the vendor might back out of the higher education market and no longer support the software.

The relations of power evident in these case studies show that there are opportunities for appropriation of the software in local contexts, but the pervasiveness of the myth that ERP engenders a best practice reality and the top-down trajectory of decision-making that supports its adoption raise the question of how effective these appropriations are in empowering those who do local design work.

Note

1. For examples see Connolly, T. and Begg, C. (2005) *Database Systems 4ᵗʰ edit.* Addison-Wesley; Kendall, K. E. and Kendall, J. E. (2005) *Systems Analysis and Design 6ᵗʰ edit.* Pearson Prentice Hall; Skidmore, S. and Eva, M. (2004) *Introducing Systems Development* Palgrave Macmillan.

7. *Reality Check: Interactivity, Reality Television and Empowerment*

KATHY WALKER

Introduction

There are many definitions of the process of communication. Fiske provides one such definition in his 1990 text on communication studies which defines the process as 'social interaction through messages'. This explanation, like others, suggests communication is a two-way process or exchange of messages. Mass communication through broadcasting however, has been regarded for the greater part of its history, as a one to mass medium of communication, lacking in any significant level of interactivity. In this asymmetric relationship the audience has been characterised, largely, as passive observer or recipient of television programming, as responsive rather than proactive. Typically, it has been argued that audiences have little say in the way the entertainment industry disseminates cultural products other than through passive consumption (Dorfman, 2002).

The advent of new technologies which facilitate greater levels of inter-activity between the producers of television content and their audiences has given rise to widespread discussion, and in some instances wild speculation, about the potential of interactive television to liberate audiences from the constraints of passive viewing. The liberating power of interactivity is expounded in terms of how 'it offers audiences a way out of the passivity and artificiality of our mainstream fictional entertainment' and a 'sense of liberation from old economic and cultural forms' (Dorfman, 2002). 'The red button empowers our audiences' it is claimed (Grade, 2004). This discourse of empowerment echoes the claims made about earlier innova-tions in communication technologies, such as the video camera and VCR

(Video Cassette Recorder), and focuses on the emancipative powers of interactivity.

This chapter examines the discourse around new communication technologies and the debates they raise about issues of control, democratisation of the media and the empowerment of individuals. It challenges the myths of empowerment and democratisation which are features of some accounts of the potentials of new communication technologies, exploring both the historical development of earlier technologies, including video and cable, and the new forms of digital interactive television or DiTV. The chapter will examine the level and nature of interactivity in television broadcasting initially addressing the technical facilities of the new digital systems which allow and even prompt some degree of two-way connection between broadcasters and audiences. Interactivity at this level might involve the use of the EPG (Electronic Programme Guide) and remote control for the menu-driven selection of television programmes, different camera angles for sporting events, or the purchase of goods from shopping channels. Discussion will then focus on exploring the dimensions of interactivity incorporated within new programme formats, particularly reality television programmes such as *Big Brother*, which, it is argued, can 'turn a passive experience into an active one' (Bazalgette, 2001), and have become the focus for claims about the great democratising promise of interactivity. If, as others have argued, information is the distinctive feature of the post-industrial society, 'interactivity' has surely become the compelling buzz-word of the new digital society. This chapter examines whose interests this vision of society serves.

The myths of empowerment and democratic participation in television

The myths relating to new media and communication technologies are underpinned by three major claims. First, that each form of communication technology is in some way new and unique and, what might be called, a break with history. As Mosco argues 'one generation after another has renewed the belief that, whatever was said about earlier technologies, the latest one will fulfill a radical and revolutionary promise' (Mosco, 2005: 8). However, technological innovation and phenomena such as convergence[1] can only be understood by analysis of the historical development of media systems. As Armes points out the development of video technology can only be properly understood if we 'see video within an overall history of sound and image reproduction which stresses the interconnection between the various systems' (Armes, 1988: 11). Similarly, Garnham (1990) argues that we need to focus on the development of media technologies as part of a historical continuum, particularly in terms of the relationship between their development and the development of consumer capitalism. The emphasis on the new and unique tends to blind us to the often strategic development of new media technologies by the multimedia entertainment industries. In terms of the Internet McChesney argues that this has meant 'most of the analysis of the future of the Internet has been imprisoned by its tendency to view the Internet as a unique and unprecedented new thing, rather than as part of a historical process and as a logical extension of the corporate media and communication system' (McChesney, 1999: 8). As the Frankfurt School argued, where profits are concerned the 'culture industries' are very good at producing essentially the same thing over and over again disguised as something new and desirable, and new technologies facilitate this process. 'What parades as progress in the culture industry, as the incessantly new which it offers up, remains the disguise for eternal sameness' (Adorno, 1975, cited in Marris and Thornham, 1996: 25).

Second, this identification of each innovation as new and unique is invariably linked to claims and hyperbole about the potential of these technologies for change. Again this is connected to the association of any new technology with notions of progress and reform. The potentials claimed for each new medium from video, to satellite and cable distribution, and the new digital systems have tended to focus on their capacity to enable audiences to participate more actively in the process of consuming television and to create a more open and egalitarian media. Providing the audience or consumer with more choice and control over the media products they consume has been a key claim for both home video cassette equipment and the multi-channel television distribution systems of cable and satellite. But it is obvious that these very real potentials are conceptualised by the media sector within a commercial, market-orientated environment. Choice for audiences has become a commercial imperative in the drive to increase consumption of television content and to encourage the take-up of new subscription and transactional-based television services. The claim by commercial media corporations to provide greater choice for audiences has become a means of legitimating the expansion of new media technologies and the basis for lobbying vigorously for deregulation of public service broadcasting (Walker, 2000).

The final strand of argument or element of the discourse applied to new media technologies is the myth of empowerment or, as Garnham argues, the 'myth of democratic participation in television'. Central to this is the claim that they provide an opportunity to challenge the dominance of existing structures of production, distribution and content of media products by widening access. The specific technical features of each innovation are said to open up a relevant sector of the media to new entrants, be they alternative producers, community groups or members of society more generally. Mosco makes a similar point in connection with the Internet. 'Computer networks offer relatively inexpensive access, making possible a primary feature of democracy, that the tools necessary for empowerment are equally available to all' (Mosco, 2005: 31). This vision is a very powerful tool in the dissemination of these technologies. As McChesney emphasises 'In the 1990s a new argument has emerged, the effect of which is to suggest that we have no reason to be concerned about concentrated corporate control and hypercommercialisation of the media. This is the notion that the Internet, or, more broadly digital communication networks will set us free' (McChesney, 1999: 119). It is important to note however, that when we analyse the history of technological innovation in the media sector 'every modern electronic media technology has spawned similar notions' (ibid.) and it can be argued that each new technology has the potential to both reinforce the structure and operation of an industry as well as to challenge it (Garnham, 1990; Hughes, 1990).

When discussing radio and television Mosco argues 'the irony, it appears, is that, as these once-new technologies lost their luster, gave up promises of contributing to world peace, and withdrew into the woodwork, they gained a power that continues to resonate in the world' (Mosco, 2005: 2). Whilst this is undoubtedly true and we can recognise the pervasiveness and influence of these technologies on the cultural and political life of society as a whole, access to and control over these technologies and their capacity to influence remains firmly with the corporations and media institutions. The initial period whilst technology is in its formative stage, and can be seen to challenge a particular media sector by opening it up to greater access and participation, is therefore relatively short-lived. In many cases the industry sector responds to the challenge or 'threat', as it may be seen, by integrating the technology into the existing structure of commodification and distribution (as in the case of audio and video tape) or by utilising the myth or discourse of choice and empowerment as promotional hype. As Armes reminds us 'the existing structures have proved remarkably resilient and capable of controlling the flood of technological innovation' (Armes, 1988: 73).

This chapter will now move on to explore the ways in which certain technologies have given rise to the myths of empowerment and democratic participation in television and to assess the extent to which they have been successful in fulfilling the democratising agenda. It will examine the ways in which the media sector has become adept in utilising the discourse of access, participation, choice, and more recently interactivity, as the means of building and consolidating new audiences whilst retaining full control of the media itself. It will argue that both programming and more significant changes to broadcasting structures are marketed by sustaining belief in these qualities. Mosco points out in relation to new computer communication technologies that 'guarantees of instantaneous worldwide communication, of a genuine global village, are in essence promises of a new sense of community and of widespread popular empowerment' (Mosco, 2005). The potential of interactivity through our televisions promises audiences access to a similar sense of community and democratic communication. This chapter will discuss the ways reality TV programmes, such as *Big Brother*, utilise the promise of interactivity to create their own pseudo-community from a viewing audience linked to on-screen characters by the belief that they can play some role in shaping the lives of the participants and deciding the outcome of the programme, who stays or who is evicted from 'the house', 'the island' or 'the game'.

Video technology – A new dawn in television viewing?

Although magnetic video tape was originally developed for industry purposes for the recording and time-shifting of broadcasting output the subsequent development and diffusion of home video was one of the most rapid of all media technologies. The technical capabilities of the VCR to fast-forward, rewind and play both pre-recorded tapes and record off-air broadcast material, were strong selling points and confirm that the cost-benefit ratio is very important in the take-up of new media technologies. By 1994 penetration rates in the United Kingdom were put at 83 per cent for all homes and 94 per cent for homes with children (ITC cited in *Screen Digest*, June 1995). Innovations in video recording and the development of the portable video camera quickly became the focus of debates about the role of technology in changing patterns of production and consumption of broadcast media. Some 30 years after the launch of the VCR technology, it is clear that in terms of the consumption of televisual programming, video recording equipment has provided the viewer with some increased control over when, how and what they watch on their televisions. In particular, the ability to record or time-shift broadcast programming has broken the control of programme schedulers giving viewers greater flexibility in managing their own viewing schedules.

However, claims that the new video format would provide a means of substantially increasing the diversity and range of programming to consumers are easier to challenge. Initially, prominent Hollywood film studios Walt Disney and Universal Studios, who recognised the challenge video recording posed for the film and broadcast industry, filed suits against Sony in 1976 claiming that its time-shift device and the advertising for it, was an 'incitement to breach of copyright'. However, the studios rapidly identified that video would give a new commercial lease of life to both old and new films and soon began to focus in earnest on the huge potential market for video software. Wasko indicates that whilst in the early 1980s it was possible for new players to be part of the home video industry, by the 1990s the field had narrowed, with a few key players, the Hollywood majors, dominating this area of the industry. The vast bulk of material available for rent or sale via the video format has continued to be dominated by these same Hollywood studios. Video distribution became integral to strategies for media synergy and merchandising and Hollywood's adoption of this new means of distribution became part of wider moves to integration and consolidation in the entertainment industry. The range of media content available is further constrained by the range and diversity of video material available through retail distributors. Local distribution from small independent video stores in the 1980s was rapidly replaced by larger chains and by the mid-1990s the video rental sector was dominated by Blockbuster Entertainment Group, the leading worldwide owner, operator and franchiser of videocassette rental and sales stores (and until 2004 part of Viacom International Inc.).

However, it was the development of the Sony Video Rover Portapak camera which primarily gave rise to a new perception of video as a tool for change. Accessible and relatively inexpensive technology prompted Armes to argue 'video is a key continuation of this democratising tradition, as a system which allows personal recording and creative production as well as the consumption of pre-recorded, pre-packaged material' (Armes, 1988: 74). The video camera held out the possibility of greater democratisation of the media by providing a new and more accessible means of production. It was seen as a way for ordinary people and community groups to become involved in the communication process and to make their own television. In the same way that the early printing press was seen as a means of circulating ideas, information and opinion which challenged the views of the establishment, access to video technology was perceived by some as a way of opening up the television sector to new and alternative voices. It was taken up in this way by a number of community, political and artistic counter-cultural groups as a tool for social investigation and means of gaining democratic access to the media. There are clear examples of video technology being used in this manner. The Cole for Dole tapes produced in 1984 presented the British miners' dispute through the eyes of the National Union of Mine Workers with 30,000 copies distributed through a structure which bypassed television (Hughes, 1990). Similarly, video was used by community action groups in Nicaragua and by Aborigines to claim their land rights in Australia.

The local cable television projects of the 1970s, although relatively short-lived in the United Kingdom, developed initiatives to bring alternative, locally produced programming to their audiences by combining video production with the alternative distribution medium of cable television.

Swindon Viewpoint, for example, which was launched in 1973, was one of the earliest experiments in community cable or public access television, which aimed to enable local people to produce programming for and about their local communities. The community video scene was also very active in the United Kingdom in the 1980s with video workshops such as Albany Video, in south London, and Chapter Video in Cardiff providing both access to, and training in, the new technology. Channel 4, with its public service broadcasting remit to support new and innovative forms of programming not generally found on existing broadcasting channels, provided a number of these workshops with funding and opportunities for broadcast through its workshop agreements. However, the more commercially competitive television era ushered in by the 1990 Broadcasting Act was to severely undermine opportunities for community video projects. The new funding arrangements introduced for Channel 4 and the subsequent demise of the workshop agreements and the introduction of one-off funding for individual productions, undermined the access, training and innovative nature of the earlier community projects. In addition, although the White Paper which preceded the Broadcasting Act expressed the view that local delivery services would provide exciting opportunities for small-scale television at city and inner-city levels, and for groups who felt that their interests were not adequately catered for under the existing broadcasting system, the Act itself removed the obligation on the newly franchised cable companies to provide local programming. Whilst most successful cable franchise bids therefore expressed commitment to providing some local programming it was clear that market forces would be the main determinant of this provision. Although the Broadcasting Act did introduce the 25 per cent rule which required all existing broadcasters to commission this percentage of their programmes from the independent sector, the increasingly competitive and market-driven nature of the new broadcasting environment favoured the large commercial independent production companies[2] rather than small independent production companies and community groups. A report commissioned by the BBC in 2005 shows that 30 per cent of the revenue expended by UK broadcasters on qualifying independent production in that year was generated by just two independent production companies, Endemol UK and All3Media, with seven other companies dominating commissions in the sector (BBC, 2005). Research by PACT indicates that 27 per cent of companies reported that they were unable to compete for commissions most of the time because of the small number of big companies dominating the market and 34 per cent complained of 'commissioners using preferred suppliers' and a 'lack of creative risk-taking' in the type of programming commissioned (Pollard E., Barkworth R., Sheppard E. and Tamkin P., 2005).

    Although the technologies of video and cable have largely failed to deliver the levels of access and local and community-orientated programming suggested in their early stages of development, it could be argued that their real legacy lies in generating awareness of the very real pleasure gained by audiences from seeing 'ordinary' or 'real' people and their lives on television. Early access television experiments, such as Swindon Viewpoint in the 1970s and the Waddington Television Village experiment in the 1980s demonstrated the huge popularity of local news and lifestyle programmes made for, by, and about local people. However, when Hollywood films *The Truman Show* and Ed TV were released it was hard to imagine the extent to which these fictional dramas would foreshadow audiences' and broadcasters' fascination with 'reality television', a new genre of programming monitoring the lives and experiences of 'real' people in much the same manner as these fictional characters. Nevertheless, since the 1980s, audiences in the United Kingdom, the United States and Europe have experienced the emergence of a wave of new programming types ranging from docusoaps, to hybridised formats such as *Big Brother, Castaway 2000* and *Survivor,* which merge social observation with game-show elements. It is interesting to note that both these genres use video technology to deliver a veneer of authenticity to the representation of people's lives whilst the production companies remain firmly in control. Video has evolved from facilitator in the production process to manipulator in the surveillance process. This chapter will now examine reality TV programming in more detail focusing particularly on the way that interactivity is utilised to encompass a participatory dimension to the programmes, exemplified in its most direct form by the voting process and indirectly as part of a wider process of cultural exchange beyond direct audience engagement with the programme.

Interactivity and DiTV (The democratic dream of the Information Age?)

In the early days of computing, interactivity referred to systems and software which allowed or prompted users to make a response by providing alternative choices for pursuing a particular choice of action from a range of predetermined menu choices (Cawkell, 1996). With the development of the Internet and broadband links the modus operandi remains the same but the concept of interactivity has expanded to include a far wider range of online applications which respond to user activity such as video-enriched information services, search engines, chat rooms and discussion forums. The change from analogue to digital format in television delivery systems, and the ability to provide reverse bandwidth or return pathway, has enabled television to begin to develop similar two-way interactive systems and enhanced services. In practice however the different digital TV platforms in the United Kingdom offer varying degrees of return pathway and subsequently different levels of interactive functionality distinguished as either interactive services or enhanced services. The broadband cable network (Virgin Media) is the only platform with integrated reverse bandwidth and therefore permanent return pathway. Digital satellite (BSkyB) has software embedded in its set-top box which provides a return channel for interactive services via a telecom link, whilst free-to-air Digital Terrestrial Television (Freeview) currently has no usable return pathway.

Technically therefore it is only cable and satellite which can utilise the return pathway to provide interactive services and it is the Electronic Programme Guide (EPG) which is integrated into each set-top box which controls the type of interactivity that is possible. The BSkyB and Virgin Media set-top boxes manage the encryption of and subscription to their pay channels and the EPG and the 'red interactive button' allow audiences to access commercial options such as video-on-demand, home shopping, pay-per-view and real-time engagement with competition and gaming channels. Although Digital Terrestrial Television currently has no return pathway the software in the EPG and the 'red interactive button' are still capable of offering viewers a more limited level of interactivity or *enhanced services*. This includes channel selection (which allows audiences to scan the range of channels on offer and their content), access to links and additional enhanced content associated with news coverage, sports events and home shopping channels, and the selection of particular views or coverage of sports events.[3] However, applications and programming which invites a response can only be accessed by means of alternative technologies. Currently therefore there are three principal forms of TV-related interactivity in the United Kingdom: via TV remote control (i.e. red button interactivity), via phone (the notorious premium phone lines) and via the web.

Broadcasters' responses to the potentials of interactivity have varied. BSkyB has made interactivity a cornerstone of its digital sports platform and its interactive services are a major strand of its advertising campaign for new subscribers (Walker, 2000: 62). Services, however, focus largely on commercial applications such as gaming, live betting, email and SMS (Short Messaging Service). Interactive applications and broadband services have also become the battleground of its current cut-throat competition for subscribers with Virgin Media. The BBC as part of its public service remit has probably done the most to develop *programming* which realises the potential of interactivity for the benefit of its audiences. Interactive features are embedded in their news and sports coverage, including major events such as the 2004 Barcelona Olympics, current affairs programmes such as *Asylum Day* (Macdonald, 2007) and entertainment genres such as *Test the Nation*. However, as William Cooper, ex-Head of New Media Operations at the BBC points out 'progress is painfully protracted and there is simply not enough real innovation. Producers are sticking with the same tried and tested formats and seem reluctant to push the boundaries. The same applications are being recycled and the rate of change remains slow' (Williams, 2005).

The use of interactivity within other forms of television programming has been relatively limited. *Spooks* (BBC), *Dubplate Drama* (Channel 4) and an amalgamated edition of *Casualty* and *Holby City* (BBC) have represented recent attempts to apply interactivity to drama series, but successes have been few and far between. As Williams (2005) argues 'creating such successful interactive programmes is difficult and expensive'. One format however with which the term interactivity has become synonymous is reality television programming where it is argued that 'the appeal of the real is inseparable from the promise of interactivity: that ordinary people have a chance to participate in creating the shows they watch and the products they buy, that the uniformity of mass society has been undone' (Andrejevic cited in Dorfman, 2002). *Big Brother* in particular has been hailed, not surprisingly, by its broadcaster Channel 4 as 'the most innovative entertainment concept ever' (Channel 4, 2000), and by Peter Bazalgette, who brought the show to UK television screens, as 'the most significant TV show of the last decade' (Bazalgette, 2005). It has come to symbolise what Dorfman (2002) has called that 'great democratic experiment we call the Digital Revolution'. Hyperbole aside, it has been argued that the programme was a turning point because of the new way it presented for audiences to interact with the programme. Curran and Seaton (2003) have suggested that all persuasive mythologies connect to an element of truth and this chapter will now move on to explore ways of explaining the popularity of this programming with audiences by examining the participatory dimension of the format and the myths that surround it.

Audience viewing figures for programmes such as *Big Brother* here and around the world show that the format has been immensely popular with audiences. 69 per cent (38 million viewers) of the population as a whole and 79 per cent of the elusive young 16 to 34-year-old audience segment, watched UK *Big Brother* 1 at least once and 9.5 million watched the final programme[4] (Channel 4, 2000). In the final phone poll, 7.4 million people voted making it the biggest ever UK televote (Channel 4, 2000). The involvement of new media and the promise of interactivity do appear to have enhanced the popularity of the format in a number of ways: by fostering audience identification with participants; by developing a sense of community amongst audiences; by facilitating an illusion of audience participation; and by enabling involvement in real-time.

As argued above, the very real pleasure gained by audiences from seeing 'ordinary' or 'real' people on television had been identified as early as the 1970s from the popularity of cable access television experiments and enter-tainment programmes such as *Candid Camera*. More recent programmes including *You've Been Framed* and the *Real Holiday Show* continued the trend of 'karaoke television' (Robins, 1996) whilst pseudo documentary-style programmes including real CCTV video footage, such as *Rescue 911* and *Americas's Most Wanted* (US) and *Police, Camera, Action* (ITV), *Police Stop!* (ITV), *Street Wars* (Sky 3), have proved similarly popular with audiences. Techniques familiar from the world of drama, short action-packed sequences and a series of emotional impacts, were used in the latter genre to draw audiences in and collapse the distance between image and viewer. Docu-soaps such as *The Cruise, Friday Night Fever, Lakesiders, Holiday Reps* and *Paddington Green*[5] built on this acknowledged popularity of real-people on television in the 1990s. Utilising fly-on-the-wall documentary techniques they also drew strongly on the formula of the soap opera; strong characters and ordinary routine punctuated by dramatic events, to encourage audience identification and involvement (Kilborn and Izod, 1997). At least nine new docu-soap series were developed by terrestrial channels in the United Kingdom in 1998 alone, resulting in the sheer saturation of this genre, and giving rise to the hybridised version of reality television with a game-show element epitomised by *Big Brother*.

Big Brother uses this same recognised popularity of real people on television and many of the techniques identified from previous programme genres to encourage audience identification with the programme participants or 'house mates'. 'The programme replicates the main theme of fictional TV drama and soap opera focusing on daily life, relationships, emotional exchanges and human interactions. As research for the European Convention on Transfrontier Television suggests 'everything they show is "so usual" and "common" that audience identification can be easily prompted' (Sampedro, 2001). Similarly, Bazalgette (2001) argues that part of the success of *Big Brother* is due to the way the programme has been embedded in the television schedules across the week, fostering familiarity and identification with the characters in the same way as those in soap operas. Considerable time is spent in selecting appropriate participants who will prompt empathetic or possibly hostile responses from the audience. Activities and tasks are arranged which bring out the worst or best in these characters and provide the spectacle through which audiences are encouraged to make an emotional investment in the characters. It is the '"meaningful" stories of love, friendship, competition or hate among the participants' (Sampedro, 2001), 'the sheer humanity of it' which encourages audiences to 'interact with the people in the programme, nice or nasty' (Bazalgette, 2001). In other words audiences care enough to want to participate. The structure of the programme produces an emotional commitment which prompts many to respond to the invitation to vote and possibly influence the outcome of the contest.

The process of emotional commitment and identification with the house-mates also underpins the concept of 'community' generated by programmes such as *Big Brother*. Although television watching has been perceived as an essentially private and individualised activity, broadcasters, and public service broadcasters in particular, have traditionally sought to extend our sense of community by constructing audiences for public, civic and sporting events amongst others. The long-term success of soap operas such as *Eastenders* and *Coronation Street* is also partly attributed to what has become known as the water-cooler effect: the tendency for audiences to gain pleasure from discussing the events, characters and plots of these popular programmes outside of the immediate viewing experience particularly during coffee or lunch-breaks at work but also at home and on nights-out. The fragmented nature of current audience viewing patterns in the multi-channel television environment mean that programmes which create this water-cooler effect are commercially lucrative in terms of increasing audience ratings and advertising revenue. Producers of *Big Brother* aware of the potential influence audience communities could play in shaping levels of engagement with the programme and its outcome, encouraged the development of audience communities by providing live Internet streaming of video footage, SMS text message updates to subscribers' mobile phones, and news, fanzines and (in Spain) tele-conferences between the participants and the public via the Internet. Virtual communities, not bounded by physical, face-to-face interactions, underpinned the 'participatory' dimension of the programme format, exemplified in its most direct form by the voting process and indirectly as part of the wider process of cultural exchange beyond direct audience engagement with the programme. As Sampedro argues, *Gran Hermano*, the Spanish version of *Big Brother*, 'allowed many Spanish people to enjoy an "elective community" while following groups or sub-groups of participants, their families and other fans together with the programme makers' (Sampedro, 2001).

The fact that audiences could engage with the programme live, in 'real-time', via the television, the Internet and SMS appears to have increased the level of identification and sense of community felt by audiences. Thus, feeling part of the action, whether simply watching the unfolding coverage day-by-day or in some cases attending the weekly eviction shows, was stimulated by the opportunity to engage with the programme by taking part in the voting process. It was argued that the simple act of voting participants out of the house changed the role of the viewer from one of 'spectator' to one of 'participant', liberating audiences from the constraints imposed by broadcasters and media institutions in traditional programming. This chapter will now move on to address whether reality television and the pseudo-interactivity it provides is really evidence of audience empowerment and the gradual democratisation of the media.

Myths of empowerment

Interactivity, like other technological innovations, is frequently portrayed as new and unique with the potential to bring about far-reaching change in the television sector. As Curran and Seaton explain,

> The theorists of interactivity were very clear that this was absolutely novel, engrossing – and very superior to old, passive spectatorship. They dismissed any idea that when you look at a picture and are delighted by it, or read a novel and find yourself transported into the mind and understanding of characters within it, watch a drama and feel involved in the outcome of the story, or contemplate news with apprehension, you are altered by the experience. 'Interactivity' was, it was claimed, revolutionary. (Curran and Seaton, 2003: 232)

As we have seen from earlier discussion, the promise of interactivity via remote control and telephone voting may heighten the sense of audience involvement and participation and make programmes such as *Big Brother* more meaningful for their audiences. However, the format also relies on the very real attraction of ordinary people on television, demonstrated over a long period of television output, and incorporates a number of techniques for engaging audiences which are familiar from other genres of programming. Sampedro (2001) argues that 'far from being real innovation, the *Big Brother* is a mixture of older types of programmes with tested success in commercial TV. The difference lies in the full incorporation of the public through the new technologies into a process of commercialisation of audience participation.'

The level and extent of interactivity is also questionable. The discourse of empowerment suggests that interactivity, most directly the voting process in the case of *Big Brother*, puts the outcome of the programme in the hands of the audience who are participants in creating the media text. However, this is deceptive since audience reaction is shaped by the selection and editing of material from the number of cameras available and remains within the authorship and control of programme producers. Audience members can only vote for the house-mates identified for eviction which in many ways reflects the inherent passivity of programmed interactivity found in other areas of computer applications. The process of nominating house-mates for eviction is also shaped by events within the house which are themselves manipulated by the structures and tasks imposed on the house-mates. Far from being empowered both audiences and participants of such programmes are more often manipulated to create the most dramatic and provocative scenes to generate controversy and increase viewing figures.

Although five million votes were cast in 2001 via the remote controls of BSkyB's digital service (Bazalgette, 2001), for most audience members voting or interaction took place through premium phone lines. In 2004 about 15 million calls were made to the eviction line for *Big Brother* which 'amounts to gross revenues of £3.8 million and that's quite a way short of revenues from the record 22.7 million votes it attracted two years ago [in 2002]' (Vass, 2004).[6] Although this revenue is split between the broadcaster and the service provider, Vass points out that it is not surprising that TV executives are looking at how best to grow the interactive market, with sport and drama being two more areas that are currently being closely considered for development by DiTV (ibid.). Williams similarly argues 'Interactive television just seems too hard, and the barriers to entry are still too high, for many producers. They are turning instead to premium rate telephony and mobile phone services, which can potentially provide an immediate revenue return' (Williams, 2005).

The technologically determinist basis of the empowerment discourse is therefore challenged by exploring the economic determinants underpinning the surge of reality television programmes in the television schedules and the limitations of the interactivity which they provide. So-called interactivity for viewers represents huge profits for the producers of these programmes. During a period of advertising downturn and declining subscription income[7] it is not surprising that broadcasters motivated to find additional sources of revenue should loudly proclaim the benefits of so-called audience interaction which generates such healthy streams of additional income. The level of participation is very limited, compared to the technological potential, and it is focused almost exclusively to the benefit of the media organisations (Sampedro, 2001). In addition, the production values for reality television programmes such as *Big Brother* are relatively cheap, with 'real' people costing far less than professionals to employ, and no scripts, costumes or other trappings associated with other genres of television production. Even proponents of the genre have conceded that 'reality television in its many forms, whether a docu-soap or a game show, can be made relatively cheaply – for between 60 and 120 thousand pounds an hour. A comedy show can easily cost 400 thousand pounds an hour, and good television drama typically costs more – perhaps a half of a million pounds' (Bazalgette, 2001).

Conclusions

The ability to provide interactive or two-way communication via the television set has been in development for over a decade. McLuhan (1964) has argued that in some ways, irrespective of the content they carry communication technologies' impact on society in the way that they extend our senses, thoughts and perceptions. The rapid development of the Internet with its seemingly in-built capacity for two-way communication has added pressure on television systems to offer a similar level of interactivity. To some extent this has been reflected in society at large: in the same way that the development of radio and recorded music technologies (the phonograph and gramophone) set up expectations for sound-on-film (the talkies) in the era of silent film, the availability of two-way communication or interactivity in other areas of leisure activity, the Internet, computer games and so on have set up an expectation of greater interactivity via our television. Commercial interests have endeavoured to introduce interactive television to the mass market for some years but as industry research shows the conceptualisation of this technology simply as another means of revenue generation poses key challenges to the potential development of interactivity including 'lack of audience interest' and 'the challenge of making the transmission from "viewers" to "customers"' (Steelside Solutions Ltd, 2003). Although the United Kingdom has the highest penetration of digital television in the world 'the same applications are being recycled and the rate of change remains relatively slow' (Williams, 2005).

One of the myths of the new television environment however, is that viewers are increasingly able to engage actively with the programmes they watch, shaping plots or, deciding the fate of programme participants. The discourse of empowerment and democratisation fails to acknowledge the relatively limited scope of interactivity on offer to audiences, usually nothing more than responding to alternative choices from a range of predetermined options. The term interactivity is enthusiastically coined to describe the voting process available to audiences of reality television programmes such as *Big Brother*. However the producers of these reality TV programmes remain firmly in control and the real people involved in the production or engaged interactively as audience members have very little real influence over the programme's outcome. Rather than empowerment, an illusion of participation through interactivity is being utilised to underpin an increasingly voyeuristic mode of television which serves the increasingly competitive demands of the television market. These can be identified as the needs to effectively target an elusive young television audience, to cut production costs, and to maximise an increasingly profitable stream of revenue genera-tion. With increasing competition in the television environment and the downward pressure on established revenue models, more innovative and truly interactive programming is unlikely to be developed when relatively inexpensive genres such as *Big Brother* perpetuate the myth of interactivity to generate lucrative additional revenue streams from premium phone line voting. As Hughes has argued 'particular changes in communications technology aren't random and inevitable, but are part of general changes in the ownership and control of major sections of the national and international economy' (Hughes, 1990: 66). The limited availability of interactivity in pro-grammes such as *Big Brother* speaks more about increased commercialisation and competition in the television sector than genuine interactivity and audience empowerment.

Notes

1. The term convergence has a wealth of definitions but is generally understood to mean the coming together of computing, telecommunications and the media, leading to the structural integration of industries which provide transmission or delivery media and those which generate content. It also refers to formerly distinct media forms or products coming together on a single multipurpose digital platform. Convergence occurs at the level of technology, product and sector.
2. Including for instance Thames Television which had lost its ITV franchise as a result of the financial bidding process introduced by the Act.
3. The number of channels available through digital television simply enables the system to switch between several separate channels running inter-related programmes thereby giving the illusion of interactivity.
4. In comparison the two Spanish editions of *Big Brother*, *Gran Hermano*, achieved an audience of 69 per cent of the population in 2000 and of 60 per cent in 2001. A European poll shows that around 61 per cent of Spaniards admitted watching the programme at least once (Sampedro, 2001).
5. In addition to the above these included *Pleasure Beach, Hotel, Airport, Vets, Vets-in-Practice, Blushing Brides, Love Town (Gretna Green), Animal Police, The Builders.*
6. In Spain 2,300 new telephone lines were established 'so that the public could vote for the winner of the programme. Each call was charged 136 ptas per minute. The official web site also enabled tele-conferences between the participants and the public, and it charged 61 ptas per minute to each cyber-visitor' (Sampedro, 2001).
7. Channel 4's previously encrypted subscription channels E4 and FilmFour were launched free-to-air in 2005 and 2006 respectively.

8. Myths, Crimes and Videotape

Delia Langstone

Introduction

The widespread use of Closed Circuit Television (CCTV) cameras in the United Kingdom means that large numbers of people who live in an urban environment are under almost constant surveillance. Many advocates of CCTV call it the 'fifth utility' (Graham 1998). This viewpoint places surveillance as an essential resource comparable to gas, water, electricity and telecommunications.

The political rhetoric surrounding CCTV attributes great things to its ability to prevent, deter and detect crime and this has resulted in a flood of funding for cameras. There is also a massive 'push' from the private sector promoting new technologies in what is now a huge industry. The reporting of CCTV in the media nearly always concentrates on its benefits. Those who take a contrary view are dismissed as the 'civil liberties' lobby or people with something to hide. News coverage utilising CCTV footage of events has enhanced the one-dimensional viewpoint of CCTV as the fifth utility and has reinforced the set of myths concerning its power. For example, the abduction of Jamie Bulger was described by Norris, McCahill and Wood as being 'replayed night after night on the national news, achieving an iconic status in the subsequent moral panic about youth crime' (2004: 111). News coverage of the London bombings on 7 July 2005 achieved a similar status in relation to terrorism.

This chapter starts with a brief description of the introduction, funding, uses and technological capability of CCTV cameras. It then seeks to compare the popular image of CCTV with reality. The myths of CCTV can be attributed to political rhetoric and to the technology being regarded as a 'silver bullet' solution to all types of crime. Next it discusses the role of the media and the many other agencies involved in the perpetuation of the myth.

It looks at the role of CCTV and the neo-liberal objectives of regeneration and the place of CCTV as a key part of the strategy to attract inward investment. The central themes, however, are evaluation of the effects of CCTV on crime and the perception of the threat of crime now known as 'fear of crime'. It also raises the question of whether CCTV has the potential in this setting to be a divisive technology. To achieve this I have chosen to look at a London borough which was an early adopter of CCTV technology. I have assessed some of the claims made for the technology and criticised the methodology used to arrive at some of the conclusions. This part of the chapter is based on qualitative research utilising the borough's own reports, statistical analysis, minutes from Scrutiny Committees, press releases and my own interviews with staff concerning their borough-wide roll out of its CCTV and Facial Recognition CCTV (FRCCTV) systems.

Applications and issues surrounding CCTV/FRCCTV in the United Kingdom

Cameras have been used in shops since the 1960s and remained mainly in the private sector (and private space) until the mid-1970s when they started to be introduced in the semi-public arena of London Underground (Norris and Armstrong, 1999: 51). Cameras to monitor traffic were also introduced in London at about the same time to aid police in controlling traffic flow. Cameras were then installed at major rallying points to monitor protesters in London and mobile cameras soon followed to film public order incidents such as picket lines and football crowds (51–53). It was not until 1985 that permanent surveillance of a public space came into being with the installation of cameras on a promenade in Bournemouth (53). Gradual adoption of CCTV as a remedy to crime and enthusiastic endorsement by the Home Office resulted in 75 per cent of the Home Office budget for crime prevention being set aside for schemes utilising CCTV in public places between 1996 and 1998 (Armitage, 2002). Since then generous funding for various crime reduction initiatives such as Crime and Disorder Partnerships under the 1988 Act (CDA) has ensured an increasing density of cameras in the United Kingdom.

The United Kingdom now has the largest CCTV network anywhere in the world (Parker, 2000: 65). It is widely considered to be an essential tool in the fight against crime. It has been cited by the Home Office as a solution for social problems such as 'vandalism, drug use, drunkenness, racial harassment, sexual harassment, loitering and disorderly behaviour' (Parker, 2000: 69). The presence of cameras is widely held to be a deterrent to criminal activity, a way of effectively deploying police or security intervention to an event and, if all else fails, as a way of providing evidence after the incident. Cameras are routinely used on housing estates to identify potential troublemakers and schools have been awarded government funding to install CCTV systems. Surveillance is used in garages, railways, shops, offices, car parks, hospitals and town centres. CCTV cameras are routinely used to monitor traffic, and since 1992 they have been used to enforce red lights and log vehicle speeds both of which have led to a massive increase in the number of prosecutions for motoring offences (Norris and Armstrong, 1999: 45).

The technology is becoming increasingly sophisticated; cameras have the ability to pan, tilt and zoom (PTZ) and provide high quality images stored on a hard drive that can be accessed and searched immediately and even remotely (Parker, 2000: 74). Cameras may have movement detectors, infrared capability for night vision, they may be small enough to conceal and they may have audio links to record or issue automatic warnings to intruders. They may also be linked to databases to provide Automatic Number Plate Recognition or Facial Recognition capabilities. Databases are capable of triggering alarms in response to 'algorithmic suspicion'[1] (Davies, 1998: 270). Some may even take their images from satellites (Parker, 2000: 73–75).

Political rhetoric often results in determinist arguments surrounding the positive image of CCTV as a 'silver bullet' solution for crime and disorderly behaviour. In their report for the Home Office, Gill and Spriggs stated, 'The Home Office endorsement of CCTV further diminished the need for planners to be seen to assess CCTV critically' (2005: 64). There are often glowing endorsements of CCTV with reductions in crime figures attributed directly to the installation and subsequent effects of CCTV systems (Norris and Armstrong, 1999: 63–64). Newspapers often run headlines along the lines of 'Cameras Catch....' with little mention of the other interventions necessary that lead to an arrest. Norris and Armstrong point out that as a visual medium television and CCTV footage are a match made in heaven giving rise to the use of clips not only in programmes such as BBC *Crimewatch* but also especially formulated programmes utilising dramatic footage that shows CCTV in a powerful and positive light (67–68). Webster identifies the various agencies involved in a complex network of activity that are 'bound together by shared goals and values, with the key goal being the diffusion and operation of the technology' (245). These agencies include 'national and local politicians, central government departments, local authorities, police forces, the media, the CCTV industry and other interested groups' all adding to the positive spin given to accounts of the technology (246).

The myth of CCTV combating major crime and the reality can be very different; the resource may be used to target what, in comparison to serious crimes, may be considered petty offences such as littering and obstruction.

Cameras may be used to intervene in instances of undesirable behaviour such as underage smoking and public order transgressions (Davies, 1998: 177). Also, there is evidence that the use of CCTV may result in the disproportionate targeting and exclusion of certain sections of society, for example, ethnic minorities, youths and the homeless (Parker, 2000: 69–70). Sites of consumption such as shopping malls are monitored in order to react quickly to the threat of behaviour perceived to be deviant (Norris, McCahill and Wood, 1998: 51). In city centres cameras are targeted to create environments that attract the 'right sort' of customer with the requisite spending power and eject the undesirable: youth, the homeless, anyone who does not 'fit in' or makes others feel uncomfortable (Featherstone, 1991: 105). As a result, spaces are changing: they are no longer places where anyone may be welcome but surveilled sites of mass consumption echoing Castells' warning of 'Technological Apartheid' (Castells, 1996: 188).

Norris and others argue that there has been a general shift from the welfare state to entrepreneurialism 'as the main motif of urban action'. For urban policy this has meant that the need to attract business into areas of depression or decline may no longer be justified as a welfare initiative but based on economic need (1998: 49). Coleman and Sim take this point further in their study of Liverpool as an area of regeneration and state that CCTV is a social ordering strategy employed to enforce 'orderly regeneration' projects (2000: 624). The whole strategy is underpinned by the drive to reconstruct areas into places of security with improved quality of life in which it is a 'safe place to do business' (626). Government policy strongly influenced by neo-liberal economic values has resulted in the reinforcement of the myth of the overwhelming success of CCTV. Essentially cameras will help to make urban areas 'nice' places to live in and able to attract financial and social resources. Crucial to the drive for inward investment is the need to divest towns of visible deterrents such as vagrants, groups of youths, litter and empty premises and to create a safe, 'consumer oriented' environment (626). As Lyon puts it, 'both city and corporation work together to create an infrastructure to foster consumption and especially tourism (2001: 61). The taxpayer foots the bill for the majority of these surveillance systems, with contributions from commercial enterprises that Lyon dismisses as the 'rhetoric of private funding' (61). There is evidence that the drive to install CCTV systems can suck funding from more 'imaginative' and sometimes cheaper solutions to local problems such as effective street lighting (Brown, 1998: 218–219). CCTV can cause what is termed as a 'Chilling Effect' on activities that are perfectly legal such as trade union demonstrations (Graham, 1998: 102). Lyon similarly recognises this worrying side effect referring to it as a 'Flattening Effect' (62).

The situation with the effect on crime is not altogether clear; rather than eliminating crime there is some evidence that cameras can displace crime into adjacent areas (Skinns, 1998: 183). Critics such as Norris and Armstrong point out significant flaws in comparisons of reported crime figures (Norris and Armstrong, 1999: 64). Ditton cites a lack of attention to seasonal adjustments and 'blips' caused by other factors (Ditton, 1999: 57). Therefore, he asserts, there is also reason to question the reliability of studies carried out and conclusions made concerning the effectiveness of CCTV (18–21). They have been described in the *British Journal of Criminology* as 'post-hoc shoestring efforts by the untrained and self interested practitioner' (Pawson and Tilley, 1994: 291). In fact, Ditton points out in his study for the Scottish Office of CCTV in Glasgow, there had been very little research at all into the effects of open-street CCTV (1999: 6). Compared to closed CCTV systems such as shops and car parks, professional independent evaluations have been a rarity (4). Some of these problems have been recognised and the Home Office commissioned a review of studies and found remarkably few methodologically sound enough to be of use (Armitage 2002: 5; Welsh and Farrington 2002: 8). Another weakness around CCTV research has been to treat all CCTV systems as the same and that their application and use are therefore the same in most circumstances.

Ditton also cites lack of methodological rigour when he criticises claims for the public acceptability of CCTV and its effect upon the perception 'fear of crime'. Fear of crime is in itself a difficult concept to quantify. It appears to be unrelated to actual crime levels and frequently the act of asking questions about it in a survey has the effect of both raising awareness of crime and increasing the perception of fear of crime itself (Ditton, 1999: 17). Statistics are often based on market research style surveys which use leading questions about crime and fear of crime but fail to mention other concepts such as civil liberty issues or alternative interpretations (Graham, 1998: 104; Ditton, 1998: 224; Ditton, Short, Phillips, Norris and Armstrong, 1999: 18). In their report for the Home Office, Gill and Spriggs concluded that with regard to overall crime rates and making people feel safer CCTV had been largely unsuccessful: 'despite all this we are still reluctant to draw the simple conclusion that it failed' (2005: 61).

So it seems that much of the debate surrounding the implementation of CCTV is based on a stereotypical view of the benefits of CCTV failing to take into account the fact that CCTV systems are discrete entities with their own specific objectives and outcomes. Norris and Armstrong point out that even though CCTV coverage is near saturation point it is still far removed from the 'Big Brother' vision. They state that systems vary greatly in significant ways namely 'scope and technological sophistication, ownership, organisation and control' (Norris and Armstrong, 1999: 55). Fussey (2004) also points out that there are a considerable number of influences at local level and all of these factors make significant differences to the outcomes of CCTV systems.

Why was CCTV originally implemented in the borough studied for this research?

In 1997 the borough published its mission statement: an undertaking to ensure that by the year 2010, the borough would be an attractive major business location and a place where people would choose to live and work. Many ways of making the borough more appealing to residents, workers and visitors were investigated and embarked upon including a strategy of crime reduction and an undertaking to reduce the public perception of risk of being a victim of crime. It was decided that the provision of a state-of-the-art CCTV system would be part of the drive to reduce crime in the area and so encourage inward investment. The first CCTV system to be used in the borough was placed in the town centre in 1997. Monitoring of patterns of criminal activity soon showed that a small percentage of offenders regularly re-offended. It was also found that although criminals were initially deterred by the presence of CCTV cameras, later they were willing to risk their criminal activity being detected by a Council CCTV Operator. Further measures were decided upon and the borough undertook to develop and introduce Facial Recognition CCTV in 1998. With the introduction of more cameras in the borough another pressing imperative behind the drive to adopt FRCCTV was the desire to rationalise and control labour costs by using 'blank screen' technology to restrain the cost of monitoring an increasing number of screens.

As the Scrutiny Committee[2] report utilising information provided by the local authority's 'Director of the Environment and the Metropolitan Police' (20 March 2001) stated, the objectives of the CCTV service in the borough were to

- Allow people to live, shop, do business and travel safely
- Create a better environment
- Support regeneration
- Enhance community safety by reducing fear of crime
- Help with the detection and prevention of crime
- Facilitate in the apprehension and prosecution of offenders in relation to crime and public disorder
- Provide the police, the council and other appropriate bodies with evidence to take criminal and civil action in the courts
- Assist in traffic management
- Reduce nuisance and vandalism.

The report stated that it is a 'London Borough that has, since the closure of the docks and resultant increase in unemployment, experienced steadily increasing crime levels'; these were consistently above the London average.

Effects on crime

A CCTV review in March 2001 written for the local authority Scrutiny Committee stated that there had been no local analysis of the net overall effect on crime of CCTV in the borough. Nevertheless, the importance of CCTV was stated as being 'undeniable' with most criminal investigations utilising CCTV footage from both the borough and private systems; the police were present in the monitor room to search archive material to this end. The mobile CCTV unit was highlighted as being particularly successful while used against racial crime, but there were no statistics to support this. In general, crime had risen nationally but the borough was said to compare favourably in comparison with its 'peer group of boroughs' in London identified by the Home Office. It was thought that CCTV had made a positive contribution to this. There were no statistics available regarding the number of prosecutions brought about by CCTV, and more specifically FRCCTV evidence, nevertheless the police stated that when they had footage it usually resulted in a guilty plea. An earlier report for the Scrutiny Committee (20 February 2001) stated that where cameras had been installed there had been an overall reduction in crime by as much as 30 per cent in some areas, although some displacement of crime to side streets was acknowledged. Overall it was concluded that CCTV reduced crime.

According to an internal report, 'CCTV Statistical Analysis Period 1997–2001' written by the Local Authority Intelligence Officer at the time, crime in total was said to be 6.4 per cent lower than in non-CCTV areas. In addition, it reported differences in offences involving drug possession (60 per cent lower, robbery a 'substantial' 39 per cent lower) with non-residential burglary also down by 58 per cent, although overall residential burglary had risen steeply and deception was up from 62 offences to 105 offences. There were also reductions claimed for vehicle crime (down 61 per cent) and vandalism (down 42 per cent). The report also included statistics of specific roads within the covert camera area and of the nine roads covered; seven showed a reduction in crime and two showed an increase. When it comes to specific areas, there was an overall reduction of crime of 24.4 per cent in one of the main shopping streets and a reduction of incidences of crime from 37 to 22 (41 per cent) in another, after the introduction of covert cameras.

It must be kept in mind that it is not the aim of this chapter to dispute the statistical evidence that the borough utilised but rather to analyse the methodology of the surveys, the conclusions drawn from the evidence and the claims arising from these conclusions. Often when the evidence supported the borough's view it was expressed as a percentage and did not reveal the sample size ('n' values). When it did not support the borough's views 'real' numbers tended to be used. Looking at the data taken from the

Statistical Analysis 1997–2001 it is possible to criticise the way the data is presented. Concerning the drop in crime figures, enormous significance was given to large movements in extremely small statistical samples without mentioning the sample size. For example, the analysis concentrated on percentages: in one area robbery reduced by a 'substantial' 39 per cent. Whilst this is true, this represents a drop from 36 to 22 reported offences, a drop of 14 in total. This has been identified as a recurring problem with evaluations of this type (Armitage, 2002: 5).

The reduction in drug possession was attributed to CCTV despite the fact that no account was taken of other factors such as a possible relaxation in the attitude of the local police to cannabis possession. This reduction cannot be attributed to CCTV unless there was evidence to demonstrate how CCTV contributed to it. This is the sort of problem in the comparison and collation of crime figures that concerns Ditton, Armstrong and Norris. Interestingly these statistics on the 'staggering' drop in drug offences and vehicle crime amongst others appear in literature promoting BT's Redcare CCTV transmission cables in January 2006; this demonstrates the use of such statistics to promote goods and services (BT Redcare January 2006).

Crime activity/geographical displacement

The problem of suspected displacement of criminal activity by CCTV was highlighted in the Statistical Analysis 1997–2001. Information from a town centre analysis looked at one key town centre street and its immediate area in the period March 1997–1998 and compared it to the same period the following year after the introduction of FRCCTV. The report noted that statistics showed 'the introduction of camera systems in High Streets has had mixed effects on crime levels in neighbouring areas' reducing the chances of house burglary, whereas crimes affecting shops and businesses were displaced into the side roads, robberies remained unaffected. It was stated as being unclear whether a change in reporting by victims or displacement might be the cause of a major increase in theft from motor vehicles.

Statistics that tended to support evidence of displacement were not considered in any depth. The drop in burglaries was described as very substantial although other possible factors that may have affected this drop were not discussed. The overall impression this report made was that the data it presented was convincing enough for the reader that the potential for displacement was not a particular problem. The data indicated a strong displacement effect, especially with motor vehicle thefts and criminal damage. It should also be borne in mind that this particular section of the report was written in response to local residents' complaints that crime had risen in the non-CCTV residential area due to what they thought was a displacement effect. The report significantly failed to discuss that crimes involving deception were up from 62 to 105; a significant jump. It failed to make a connection with this displacement of the type of crime rather than just the area the crime was committed in. This is noteworthy as it may have shown that CCTV detects 'physical' crime but not 'intellectual' crime.

Public opinion on CCTV

One of the myths surrounding the introduction of CCTV is that it has virtually 100 per cent public support and only those who have 'something to hide' would object to it. The Intelligence Officer suggested that there was no resistance but if there was it would not be law-abiding citizens complaining. He and the controllers contacted for the research were genuinely mystified why anybody would object. Surveys carried out by the borough showed that as many as 92 per cent of people supported the use of CCTV in the borough. It is advisable, therefore, to look at where the statistical samples came from. The effects on crime and the way the benefits of CCTV are presented to the public is a major factor in the public's acceptance of CCTV. Answers to the only question about the cameras were as follows: (see Table 8.1).

These figures were quoted as being 9 out of 10 residents being in favour of CCTV used to monitor damage to property and general vandalism. There was less support (8 out of 10) for it being used to identify those who park illegally and drive in bus lanes. The conclusion given is that the public do support the use of CCTV. The survey also showed that White residents showed greater support for the use of CCTV for all of the above activities apart from illegal parking and driving in bus lanes. English speakers were

Table 8.1. East London Panel Survey 2 – Q19, Use of CCTV 2000

'CCTV is currently used to detect crime against people (e.g., street robberies). Do you think the use of CCTV should be expanded to identify and prosecute people doing the following'?	Yes (%)
Vandalism in car parks	93
Vandalism in parks and green spaces	91
Vandalism to parking meters	89
People who abandon cars	87
People who dump rubbish	87
Illegal parking	81
Driving in bus lanes	81
None of these/don't know	0.2

more likely to generally support the use of CCTV than people who spoke English as a second language. So the survey showed there was mixed acceptance of CCTV between ethnic groups.

The results from the above study were taken from a general public opinion style survey of the type that Jason Ditton criticised. Respondents were asked the following question: 'CCTV is currently used to detect crime against people (e.g., street robberies). Do you think the use of CCTV should be expanded to identify and prosecute people doing the following'? This was followed by a list of anti-social activities such as vandalism. This is reminiscent of Ditton's findings in which he identifies a problem with the way that questionnaires are worded. The above question from the borough survey was couched in such a way that it elicited a predictable response from the interviewee. This is not to say that this was a deliberate ploy devised by management in the borough to skew survey results. They were, however, guilty of being insufficiently rigorous in their methodology and therefore causing an unintentional bias. This was potentially serious because of the weight they subsequently gave to the findings; they were often cited by representatives of the borough in support of various arguments in favour of more cameras and in press releases. Once again these findings are taken from a very small statistical sample and are represented in percentages.

A 'Listening Day' was held in 1999. This is when other council members such as senior local government officers go out and solicit public opinion with regard to council services. Members of the public were asked to express their opinions and it was said that support for the use of CCTV had risen to 92 per cent of people surveyed. This was a study that also lacked method-ological rigour. It is difficult to imagine a representative sample of the population of the borough being interviewed during one day, in one location, by a few representatives of the borough and the police. Nor is it feasible that reliable statistics could have been collated from what amounted to an informal chat about local amenities. These results were subsequently given as much significance as a properly conducted survey and were readily quoted in order to add weight to pro-CCTV rhetoric.

The most obvious instance of opposition to CCTV/FRCCTV mentioned was the vandalism of equipment in a local market area. Representatives of the borough regarded any objection to the use of the technologies as neither a civil liberties nor a privacy issue but the desire of the criminal element to continue to ply their criminal trade. There was very little mention of any resistance to the technology in their report and there had been no survey that specifically addressed this subject. Vandalism of cameras had been a considerable problem in the market area. Equipment had been vandalised by youths (10–18 years old) who the report stated as seeing the cameras as a 'threat to their social and criminal behaviour' (Scrutiny Committee Report, February 2001).

Fear of crime

One of the stated aims of the CCTV system in the borough was to 'enhance community safety by reducing fear of crime' and the 'public perception of risk'. Consequently, this is another area that came under scrutiny. Research[3] into whether CCTV had reduced fear of crime was conducted in the form of a survey of 107 users in one town centre thoroughfare and 30 businesses and were asked 'If CCTV made them feel safer'? If we ignore the fact that this is a very small statistical sample out of a resident population at that time of approximately 235,702, we may go on to analyse the results and the way in which they are presented. From this statistical sample the following findings were given: 38 per cent of the street users felt safer and 53 per cent of businesses felt safer. However, these two figures representing approximately 41 people and 16 businesses respectively gave rise to the unqualified conclusion that 'CCTV does assist in reducing the fear of crime' (Scrutiny Committee Report, March 2001).

Crime was said to be decreasing in the borough and was also stated by its representatives to be less in comparison to surrounding boroughs according to Home Office data. So it was surprising to learn that fear of crime was still on the rise with an increase of 17 per cent on the previous year and with 56 per cent of residents saying that fear of crime was in their top three concerns. The report also stated that this concern was 12 per cent higher when compared to the rest of London. In a similar vein, fear of crime was seen to affect the quality of life of a higher proportion of residents than in a national crime survey. Even though the majority, 71 per cent felt it only had 'minimal impact' or 'moderate impact' (51 per cent and 20 per cent respectively) compared to the people who were 'greatly affected' (28 per cent). Seventy-one per cent is a large percentage and in no way supported the comment in the accompanying report that stated: 'evidence suggests that residents perceive crime to be a big problem' in this borough. The results were extracted from the general public opinion survey mentioned earlier conducted by the borough in which there was only one question concerning fear of crime. However, it remains that if we are to take the conclusions at face value, one of the stated aims of CCTV to 'reduce the fear of crime' must be deemed a failure.

A divisive technology?

As discussed earlier, CCTV can result in certain members of the public, such as ethnic minorities, youth and the homeless, suffering from infringements of privacy and freedom of association. Youths were singled out for special attention in the report written by the Intelligence Officer. Similar problems may also occur with people who are deemed to be vagrants and there is evidence of males being targeted more than females for surveillance and intervention. CCTV controllers told me that they concentrate on young males and congregating youths as they tend to be where the trouble may be. Whilst this may be understandable from their point of view and perhaps in the light of their experience this clearly echoes the findings of Ditton, Norris and Armstrong discussed earlier. I was also told that one of the aims of the introduction of CCTV was to remove vagrants from the shopping centre where they congregated. They were generally not particularly rowdy or engaging in criminal activity but just deemed undesirable in a retail location. As this chapter has discussed, CCTV has been found to be used to exclude persons known to operators irrespective of whether they were doing anything wrong and even for reasons as petty as being deemed 'scruffy' or appearing drunk. This illustrates the current 'emphasis on exclusion as (the) dominant strategy of social control' (Norris, McCahill and Wood, 2003: 46–47; Hempel and Töpfer, 2004: 8)

Conclusions

This chapter describes both the claims made by CCTV supporters and the deficiencies in some of these arguments. The aim is not to discredit CCTV but to de-mythologise it. CCTV has a purpose and gives rise to some benefits but it is not a panacea for crime, social disorder and social exclusion. Hence, it is worth reconsidering why the borough may have presented CCTV in such a positive way and how this adds to the growing body of myth surrounding the technology.

The borough firmly believed that CCTV achieved its aims in terms of crime prevention and detection and making people feel safer in their community. It supported this belief with crime statistics and results from its own surveys. However, it must be acknowledged that these surveys were based on very small statistical samples in relation to the residential population. One can reasonably pose the following questions. Were the crime statistics and survey results robust enough for the borough to justify the extent of success with crime and a safer community that it attributed to its cameras? Also, why did the borough appear to overstate the ability of CCTV to detect and solve crime rather than acknowledging that the issues were more complex in some instances? The borough was very keen to play down negative aspects of the system such as geographical and activity displacement, increases in crimes such as motor vehicle theft and the overall failure of camera presence to prevent crimes such as deception and dipping (pickpocketing). What it did acknowledge was an increase in the fear of crime; something their CCTV strategy explicitly sought to reduce.

One of the aims of the 'vision' was to make the borough a more attractive place for business and so encourage inward investment, with reduction in crime through the use of CCTV/FRCCTV as one of the key parts of achieving this. As a borough spokesman put it, one of the main objectives of CCTV was 'to support regeneration'. Clearly the use of these technologies is therefore, inextricably linked with the objectives for business within the borough and the success of the regeneration plan. The innovative nature of the technology has certainly raised the profile of the borough; understandably, it put a lot of energy into crime detection and prevention and pivotal to this was the use of a CCTV system. The borough has made a great effort to sell the system to people as can be seen by their vigorous marketing of it. Accordingly, it would be hard for them to acknowledge the possibility of a two-tier borough. The regeneration of the area may well be only partial, with places where consumers may safely shop to the benefit of businesses, and 'hotspots' of crime driven to the periphery. CCTV is often regarded as blurring the distinction between public and private domains. Perhaps what is a more apposite distinction is the creation of two distinct domains, the 'observed' and the 'unobserved', be they public or private.

The borough garnered public support for its CCTV scheme by capitalising on the climate of unquestioning belief in the technologies; it appeared at that time to be corporate policy to pursue many avenues of media interest to further promote the notion that in this borough CCTV was an omnipresent watcher of the potentially deviant citizen. Pivotal to this strategy was the promotion of the idea of its effectiveness; as the borough reports argued, word gets around about how effective FRCCTV is and that is half the battle in crime prevention. There was no acknowledgement of the shortcomings of the technology at that stage of its development. The strategy of the borough and the police was to imply that CCTV was solely responsible for any arrests made, in the sense that 'no effort is made to deny the fact' that the arrest is not related to the technology (Statistical Analysis, 1997–2001) Norris and Armstrong's assertions discussed earlier which point out that a positive spin is used in relation to CCTV were borne out by this research. The council used local and national press and a steady stream of press releases and photo opportunities to further this impression. This resulted in the public perception about the pervasiveness of the cameras and the extent of the impact in terms of crime detection and prevention. This met one of the objectives of the scheme which was to make it seem bigger and more active than it really was in the hope of increasing its effectiveness. The borough appeared to be satisfied with the level of public acceptance of the system. However, as levels of surveillance increase in the borough there is a danger that public compliance may decrease if the surveillance appears disproportionate to the need. As Sir John Smith, the former deputy commissioner of the Metropolitan Police warned, too much surveillance will ultimately 'distance the civilian from the state' (Davies, 1998: 245). There is also research that suggests that over time residents become disillusioned with CCTV when they realise it does not live up to its initial promise (Gill and Spriggs, 2005: 57–58).

The borough is to be applauded for undertaking evaluation of the impact of its CCTV operation and in some instances responding to the wishes of its residents for the installation of cameras. Also for being aware of other environmental factors such as poor street lighting, hedges, alleyways and for introducing other methods of crime control such as community wardens. However, that only served to muddy the waters further in terms of ascertaining what really impacted on crime in the borough. There can be no substitute for the utilisation of studies using a rigorous methodology from data collected over an extensive period of time. In the absence of rigorous studies, the danger of investing in expensive technology that only tackles transitory problems, fails to alleviate, or worse, exacerbates problems in the future will persist. The idea of progress via a technological fix for social problems with the promise of efficiency, infallibility and rationality is too tempting a prospect to discount and has enduring appeal. However, the truth of the matter is that the invariably complex set of social relations contributing to problems may prove impervious to what technology has to offer for many reasons:

- Technologies are not neutral; they may have uneven benefits and adversely affect certain sections of society.
- What works in one setting may not succeed in another – the context must always be taken into account.
- Not all those who oppose are Luddites, have another agenda or are potential criminals hiding something.
- Public acceptance may change over time.
- There is no such thing as a purely technical solution – the best schemes are part of a wider initiative.

CCTV is peculiar in the sense that it is reliant to a certain extent on its own mythmaking in order to bolster its perceived effect. It is understandable therefore, that this and many other boroughs were excited by the promise of CCTV and attracted by the positive spin the government put on it supported by grants and awards. The perpetuation of the myth gradually became an accepted part of the strategy to achieve both maximum impact from cameras and as justification for buying wholeheartedly into such an innovative and hitherto untested technology. Authorities would be advised to be wary when considering investment in any future 'magic bullets' and to stay immune to the hype surrounding new technologies being developed that rely even less on human agency.

Notes

1. The digitalisation of CCTV systems: the utilisation of databases and increasingly sophisticated software for the recognition of number plates, faces and suspicious behaviour enabling the creation of 'categories of suspicion' and various triggers for responses to what is deemed unacceptable behaviour.
2. Consists of the Scrutiny and Overview Committee, Scrutiny Commissions and Working Groups made up of elective non-executive and co-opted members who meet to gather evidence from officers, the Executive, service users, expert witnesses, the public and external partners to examine particular issues of local concern, policies and council services. Required by every council since 2002.
3. A university was commissioned by the local authority Corporate Strategy and Resources Scrutiny Committee to look at the impact of a specific element of its Single Regeneration Budget.

Part III Myths about Nature, Society and Biotechnology

In this section, myths about nature, society and biotechnology are explored. In earlier times, the distinction between nature and society formed the bedrock of much Enlightenment and subsequent social thinking and scientific investigation.

In Cudworth's chapter, the boundaries between these worlds are examined. In the late 1980s, Donna Haraway envisaged a future populated with cybernetic organisms. She argued that new technologies and the social forms that arise from them were reconfiguring the boundaries between humans, animals and machines so that we are becoming hybrid entities. Donna Harraway's concept of the 'posthuman' makes much of examples of close interactions between the human mind and technology: it is full of ideas of boundary crossing between humans, other animals and machines. But these kinds of technological intervention in nature are not neutral acts. Some have argued that they are shaped by various kinds of social dominations and inequalities, based on class, gender, race and postcolonial relations. Others suggest that human societies have always transformed nature by their technologies: 'nature' is, and always has been 'socially constructed'. Nature has long been used in production processes, for example, in the making of animal products such as milk, meat and eggs. The powers of animals have been subjected to progressively intensive rationalisation, automation and mechanisation in the production of commodities for human consumption.

However, Cudworth argues that while comparisons can indeed be made between computation and the lives of animals, this does not mean artificial, human and animal life are all the same. Because humans and animals are embodied, the differences between their thought processes and the way machines 'think' are fundamental. In this sense, Cudworth argues that we can distinguish between categories, while allowing space to rethink each category in its own right.

Turning to the chapter by Davis and Flowers early in the HIV/AIDS epidemic there was no treatment, no blood test for the virus, or even any explanation of what was causing AIDS. But there have been significant achievements in the medical management of HIV since HAART (*Highly Active Anti-Retroviral Treatment*) was introduced in the mid- to late 1990s, at least in countries that can afford it. There is now a dominant clinical discourse that through HAART, HIV has become a manageable chronic illness. However, HAART can have serious side effects that alter the appearance of users or lead to life threatening disorders of the lipid system. Side effects can become so serious that patients stop treatment or switch to new treatments.

Nightingale and Martin point out that the existence of a medical 'biotech revolution' – of which HAART forms part – has been widely accepted. This myth has generated expectations about significant improvements in the drug discovery process, health care and economic development that underpin a considerable amount of policymaking. But rather than producing revolutionary changes, medicinal biotechnology is following a well-established pattern of slow and incremental technology diffusion. Consequently, many expectations are wildly optimistic and over-estimate the speed and extent of change.

Finally, the chapter by Senker and Chataway suggests that biotechnologies have roles which are potentially very valuable, not least in improving public health strategies such as sanitation and vaccination. But current mythology places exaggerated expectations on the potential of science and new technologies for improving the health of the world's population. The myth that agricultural development, in general, and genetic modification, in particular, are mainly driven by the motivation of feeding the hungry is not supported by evidence. Large multinational corporations (MNCs) have driven the principal technological changes which have affected agriculture in the last several decades. Their aim has been to make profits from agriculture both in the developed world and in developing countries. A common factor in the policies they have is to secure the largest possible markets for their products regardless of the consequences for human health and happiness, and a common factor in the routes they have followed to secure this goal has involved the worldwide diffusion of monocultures. Nevertheless, there seems to be some potential for agri-biotechnology to contribute to the improvement of food availability to the hungry – for example by means of Public-Private Partnerships.

9. Nature, Culture, Technology: Myths and Inequalities in the Posthuman Zoo

ERIKA CUDWORTH

Introduction

A dominant theme in the news media and in popular culture is that new technologies have precipitated a dramatic and irreversible change in social life wherein we have moved into a new era – the 'information age' (Castells, 1996). Tales of progress, democracy and equality surround many of these developments (Mosco, 2005) yet in certain areas, particularly bio-scientific technologies, critical voices make their presence felt, and these are often dystopian (Caygill, 2000). Some critiques suggest that technological intervention in 'nature' is shaped by various kinds of social dominations and inequalities, based on class, gender, 'race' and postcolonial relations. Some also consider that the contemporary extent of human manipulation in 'natural' processes is itself questionable.

Others, however, suggest that human societies have always transformed nature by their technologies. 'Nature' is and always has been 'socially constructed' (Eder, 1996) and rather than speaking of nature and society, we might refer to 'social nature' (Blaikie, 2001), or 'naturecultures' (Haraway, 2003). In this view, there is no pure, unsullied 'nature' 'out there'. Rather, what we think of as natural and social 'things' are hybrids, composed of a fusion of 'natural' (living, non-human entities) and 'social' (human-created) objects and processes. For example, the plants grown for display in a domestic garden or for food in agriculture are not entirely natural objects, but the product of many years (decades, centuries or more) of modification and breeding by gardeners and agriculturalists. In addition, such plants and crops are grown in particular ways and for certain purposes, which reflect human desires and needs. Some of these theorists go further than this and see the fusing of technology with 'nature' as progressive, for example, agricultural biotechnology enables a future in which scarcity of resources and inequalities between different social groups are obsolete.

In the late 1980s, Donna Haraway (1991) was one of those who argued that human beings are becoming 'cyborg' as the boundaries between them, machines and non-human animals are blurred due to various kinds of technological developments. In this view, we are moving towards a 'posthuman' future in which there will be no essential differences between artificial and natural 'life'. To argue that many 'things' are hybrids composed of both social and natural elements is one thing. It is another, however, to argue that the technological intervention with or in human and animal bodies has rendered us 'cyborg', and that we have entered a 'posthuman condition'. The notion that 'we' are cyborgs and that we currently live in a condition that is posthuman is a myth. This chapter argues that there are important physical boundaries that distinguish humans and non-human animals from machines. There are also differences and social inequalities in human social relations and human relations with other animal species.

The concept of the 'posthuman', as it is currently articulated, tends to marginalise the significance of the embodied condition of our species because it makes much of examples of close interactions between the human mind and forms of technology. In much of the writing about our 'posthuman condition', our embodiment – our human situation as creatures with a concrete physical presence – is rather lost. This is not necessarily a 'new' development, and historically, prevailing Western notions of the 'human' have had a problematic relationship to human bodies themselves (Turner, 1984). What is particular to the idea of the 'posthuman' is that the distinctions between human bodily existence and computer simulation, between organisms and cybernetic mechanisms are unclear, and in the work of some theorists and researchers, such differences are no longer held to exist.

This chapter will argue that the narratives of cybernetics and artificial life embody fantasies of escape from the individual human body and from our condition of embodiment. These fantasies form part of a complex array of myths of new technologies (Mosco, 2005). The specific myths of the posthuman are that new technologies, particularly genetic engineering and cybernetics, are taking us into a world beyond the human, in which nature and technology and humanity are indistinguishable, and in which this development is seen as progressive. This chapter considers such assertions to be mythic, particularly in that together, they suggest a 'posthuman condition' that is supposedly transgressive of norms of social inequality.

Animals, humans and 'the posthuman'

'Nature', 'society', 'human', 'animal' – these are all are highly contestable terms. Increasingly it is becoming accepted in the social sciences, that nature is social and variably constructed across time, space and place (Macnaghten and Urry, 1998; Soper, 1995). Some sociologists are suggesting that sociality is not exclusively human. Both humans and certain other animal species exhibit certain common characteristics of social behaviour, and stand in social relationships to one another (Benton, 1993: 68). In addition, the 'human' is also a human invention, a social construct linked to formations of power, wherein the norm of 'humanity' became historically identified, for example, with both Western 'civilization' and technical control over nature (Anderson, 2001: 80).

Biologists such as Lynn Margulis and Dorion Sagan (1986: 214) have argued that there is 'no physiological basis for the classification of human beings into their own family'. Such taxonomy is a product of 'anthropocentric', that is, human-centered, bias. We are, they say, great apes, and the category of 'human' owes more to religious belief and cultural values than it does to zoological distinction. In philosophy, Peter Singer (1979) is well known for his controversial advocacy of 'rights' for animals, and most recently for his involvement with the 'Great Ape Project', which proposes basic moral and legal rights for 'non-human great apes' (Cavalieri and Singer, 1993). The power relations and dominant social, economic and political institutions of Western modernity have been constituted by and through constructions of social inequality, of class, race and gender. However, these social categories of difference and domination have also been crosscut by prevailing ideas about 'nature'. This has had implications for the treatment of certain categories of humans who are 'natured' and thereby seen as closer to nature or 'less civilized' (Anderson, 2001; Merchant, 1980) and has certainly impacted on non-human species of animals, many of whom are seen as means for the satisfaction of human ends. Should the 'human' be recast in a way that compromises our distinction from non-human animals, the implications would be profound.

This said animals have always been part of human cultures. Sociologists such as Keith Tester (1991) and Adrian Franklin (1999) argue that human relations with other animals tell us something about the specifics of human cultures in particular places and times. Tim Ingold (2000) notes that in some cultures, the distinctions between humanity and animality is unclear, and clear moral and legal distinction is a modern and Western phenomenon. The idea of the posthuman is bound up with such questions of distinction, for example, between the language of computer coding and the linguistic structures deployed by humans (Hayles, 1999: 279).

The 'zoo of posthumanities'

Ideas about the posthuman abound with ideas of boundary crossing between humans, other animals and machines. Judith Halberstram and Ira Livingston (1995: 3) suggest that given the social constitution of both humanity and animality, we humans are best seen as part of a 'zoo of posthumanities'. However, an understanding of social natures and the hybrid constitution of the social/natural/technological must be cognisant of the persistence of social difference and inequality. Talk of a continuum of differences between species, for example, must acknowledge that the keepers of the zoo have not always been benign. Modern societies have incorporated certain species of animal within industrial agricultural complexes in which these embodied creatures function as sources for food processing and labour power.

Kathryn Hayles (1999: 34) defines the 'human' in terms of an historical epoch associated with the Enlightenment tradition of liberal humanism, when the human individual is taken as the basis of most understandings of what it means to 'be' in the world. She claims that the 'posthuman' can be defined as an epoch in which 'computation...is taken as the ground of being'. This means that humans, non- human animals and machines are generally understood in terms of codes and signs and rather than being seen as distinctly different, they are conceptualised as (relatively) seamless things. Technologies associated with virtual reality often have the potential for full- or part-body interaction with computers. For example, through the sensations of a flight simulator or through machines which attach to/cover parts of bodies (such as the hands or eyes) as part of computer games. In such cases the boundary between human and machine is permeable: in computer gaming, the user of the machine effectively plugs the human sensory system into a direct feedback loop with the computer (Sterling, 1992; Zuboff, 1988) and both the machine and the human player enter an interaction in which both parties alter their behaviour in response to the actions of the other.

For some of those engaged with the science of cybernetics, like Hans Moravec (1988), these are prescient developments. We are in an age in which our bodies are both best understood in terms of informational codes, and in which they are in a process of physical dematerialisation as carbon life forms. Embodied minds are replaced by artificial life (AL, often seen as cybernetic silicon forms which grow, reproduce, mutate and regulate themselves) and artificial intelligence (AI, usually meaning intelligent machines). Moravec argues that we are all keenly aware of the limitations of our physical bodies and our finite presence on the planet as embodied beings. In cyberspace however, what is of overriding importance is data not physicality. In this sense, a presence in virtual reality is potentially immortal, and the implications of this are significant. Moravec sees a future that is 'postbiological', in which artificial life and intelligence have superseded the human (116–122). Such accounts of the posthuman literally assume the increasing obsolescence of the human as we either become more like machines, or are displaced by intelligent machines as the dominant life form on this planet.

Others do not share Moravec's apocalyptic view. Humans and machines co-constitute each other, and we interface with machines so often and so intensively that it is no longer meaningful to separate us from machines (Reingold, 1991). As part of this symbiosis, some consider that human pleasure and desire is expressed in everyday flirtations with the posthuman. These can be relatively dramatic – the child using the flight simulator at a science museum for the first time, for example. They can also be more mundane. For Allucquére Roseanne Stone (1995), everyday interactions such as sending and receiving e-mails, or engaging in a multi-user dungeon (MUD) question our perceptions of our embodiment and our identity.

However, the extent to which this constitutes a decisive break with the past is questionable. One might see the keys of a typewriter or a pen as an extension of one's self, similarly to a computer keyboard. I would not suggest that there is little difference in the historically specific relationship between humans and typewriters and humans and personal computers, but the conception of a completely different configuration between humans and technology is overdrawn. The integration of the human with machines, particularly computers, has been the focus of theorisations of the human condition as becoming 'cyborg'. I want to examine the myth of the cyborg in more depth, as some of the beliefs which surround it come together in a grander system of belief – a mythology of a posthuman 'condition' which some consider to be in the process of being realised.

'We are all cyborgs now' – Utopian myths of the posthuman

In the late 1980s, historian of science Donna Haraway envisioned a future populated with cybernetic organisms. She argued that new technologies and the social forms that arise from them were reconfiguring the boundaries between humans, animals and machines (1991: 165) so that we are becoming hybrid entities (150–151). The concept of the cyborg, as far as Haraway was concerned, helped to question simple and inaccurate 'unitary' constructions of social differences and inequalities such as gender, 'race' and class, and some feminists have seen it as a liberatory symbol (Alaimo, 1994; Balsamo, 1996) although not unreservedly so (Lykke, 1996; Stabile, 1994). For Haraway, the cyborg also suggested new ways of relating to the non-human (1991: 170–172) because it muddied the boundaries between the human and non-human. Some have usefully drawn on these ideas of boundary crossing and the complex qualities of difference between humans and amongst animals in debates in environmental ethics (Cheney, 1994). Such arguments draw on the ideas of Bruno Latour (1993) who considers that objects are always hybrid, composite entities, which operate within 'networks' of relationships. Latour suggests that we must understand the world around us to be composed of entities that are usually social, natural and technological all at the same time.

Haraway's notion of the cyborg pushes this understanding of the hybrid quality of life further in arguing not just that 'things' are 'assemblages', that is, composed of various influences, processes and material qualities, but that they are constantly shifting. In this liquid world, it is difficult to talk of fixed entities, like humans or 'animals'. There are two interrelated problems with this understanding of our cyborg condition. First, it cannot capture continuing relations of power and inequality in social, political and economic life, which shape the development and use of technologies. Second, it has the qualities exemplified by myths. The story of the cyborg tells us that hybridity is both ubiquitous and progressive: 'we' humans are cyborgs now, and so too are most of the objects with which we come into contact. Our future is one where cybernetic organisms are normative and desirable, for the use of technology enables us to minimise some of the harsher aspects of the 'old' human condition (such as want and poverty, illness, aging and disease). This is a powerful story, but a story it is. We can understand the complex and hybrid qualities of objects and relationships, between humans, animals and technologies, without accepting the myth that these are inevitable, progressive and undermine social inequality (Whatmore, 1999).

Laboratory games – The case of 'OncoMouse™'

The myth of the posthuman is most clearly articulated in Haraway's collection *Modest_Witness@SecondMillenium.FemaleMan™_Meets_OncoMouse™*, which explores the discourses and practices of contemporary 'technoscience' in the United States. Haraway suggests technological developments mean that we should collapse the usual conceptual distinction between humans and animals. One important example to illustrate her argument is 'OncoMouse', the world's first patented mammal, which was genetically engineered by the chemical multinational company, DuPont. Oncomouse is a research 'tool' born with cancer bearing genes, to which Haraway refers as a biotechnological 'actor' (1997: 97), suggestive of a future in which animals and plants become 'fully artifactual' (108). A focus of her discussion is the representation of animals and humans in bio-technical product advertising. Haraway describes the cyborg representation of chimpanzees, mice and rabbits as 'wonderful' (254), ignoring issues around the use of mammals in experimentation (Birke, 1994), and the gendered, natured and racialised content of the representation of these animals. Yet Haraway sees advertisements that carry gendered and racial imagery in representing humans as 'ominous' (1997: 255–265). Unable to see social domination when looking at non-human animals in biotechnological contexts, she is troubled by the similar representation of humans.

OncoMouse™ is a biotech commodity, incarcerated, programmed with disease and designed to suffer in order to help humans 'cure' cancer. Bodies, of both humans and other kinds of animals, are shaped, altered and disciplined by social structures that reflect formations of power. The relations of nature, gender and capital are vital in understanding contemporary manifestations of the technologising of 'nature'. The commodification of non-humans is the basis of Vandana Shiva's analysis of biotechnology wherein plants and animals become 'instruments for commodity production and profit maximization' (1998: 29). Globalising biotechnologies involve the 'predation' of one class, 'race', gender and species on others wherein the 'dominant local' seeks global control (105,122). Organisms and eco-systems lose bio-diversity, domestic animals suffer abuses in genetically engineered meat production, bio-technological seeds result in superweeds and superpests (37–41), agricultural communities in the global South are disempowered by the imposition of patenting restrictions and forced to import Western seed (59). (See also Chapter 12 in this volume.) Unequal power relations around class, 'race', gender and 'nature' frame Shiva's analysis, which examines both material changes in technological applications and their effects, alongside the shifting beliefs with which they are associated. Contemporary forms of hybridisation, whilst involving physical interpenetration across species, do not easily contest power, but often remain embedded in the social networks of domination.

The cyborg is a powerful myth of posthumanism. There are important differences between animals, humans and machines. Some of these are also myths, such as the crude separation in Western cultures between the species 'human' and the conflation of the incredible variety of non-humans into the category 'animal'. However, it is a form of technological reductionism that posits the inevitable march of progress of technological 'development', which is so embedded in our lives and practices that it cannot be resisted. (See also Chapter 4 in this volume.) Vincent Mosco (2005: 21) notes that 'what some have called "embodied physicality" is the unrecognised sibling of the more popular notion of virtual reality'. Embodied physicality is a sobering counter to the flights of fancy latent in the cyborg myth, and it is to this we now turn.

Physical embodiment and social inequality

The posthuman can be defined as a privileging of 'informational pattern over material instantiation' (Hayles, 1999: 2). What Hayles means here is that the posthuman worldview sees no essential differences between human bodily existence and computer simulation, between organisms and cybernetic mechanisms, as humans and other animals are seen primarily as information processing machines. This is not an entirely 'new' development. In most social theory, human beings have been seen as distinct from 'animals' and having particular value due to their capacity for rational thought. Our situation as creatures with bodies has often been ignored.

Some recent social theorising has attempted to address this bias with a focus on the human body and the notion of embodiment, but unfortunately, it has understood the body as purely social, that is, it has reduced physical bodies to social ideas. For example, according to Rosi Braidotti, the body is the

> interplay of highly constructed social and symbolic forces: it is not an essence, *let alone a biological substance.* (2002: 21, my emphasis)

Such theorising draws on the ideas of Giles Deleuze and Felix Guattari (1983) and their 'body without organs'. Here, the body is understood as a collection of symbols rather than a physical entity (Gatens, 1996). Elizabeth Grosz (1994) emphasises the fluid qualities of bodies as inscripted surfaces, that is, as objects which are given their character by the culture in which they find themselves. However, by ignoring the physical body, such theories implicitly assume the physical body is a constant, fixed phenomenon (Birke, 1999: 138), underneath the cultural imprinting with which they are so concerned.

More helpful have been theorisations that understand the ways embodiment reflects social relations of inequality. These have often drawn on the work of Michel Foucault in outlining the powerful practices that produce 'docile bodies' (Foucault, 1979: 138). We discipline our bodies; train them, to make them socially acceptable. For example, bodily modifications like dieting can be seen as gendered forms of embodiment (Bordo, 1993: 90). Pierre Bourdieu (1984: 466–467) argues that we incorporate social values and norms into our bodies as ways of talking, standing and walking, and these practices reflect social inequalities.

The body is 'simultaneously biological' and 'shaped by but irreducible to contemporary social relations' (Shilling, 2003: 182). Social theorists have critiqued understandings of bodies within the natural sciences (see Keller, 2000). For example, the mechanistic notion of a body composed of replaceable pieces of machinery exchanged through market relations (such as the international trade in replacement human organs), and across species boundaries (xenotransplantation, involving, for example, the use of pig heart valves to replace deficient human heart valves) (see Kimbrell, 1993). However, there is a shift in parts of the biosciences towards non-mechanistic understandings. Lynda Birke suggests that human and other animal bodies might be conceptualised as located in a dynamic environment with which they co-evolve. Social change (such as housing provision), the impact of war, famine and natural disasters, and different cultures of consumption (diet, for example) affect bodies physically. These social factors become embedded in the physical body as forms of information and they change the actual functioning of physical bodies (Birke, 1999: 107). In such theories, our bodies are seen to be both dynamic and self-regulating, with tissues and cells in constant flux. As Stuart Kauffman (1995) suggests, we may appear 'fully' formed in our human embodiment but our bodies are not fixed. They are complex bundles of possibility, dynamically engaged with physical and social systems.

So, human bodies embody layers of complex social relations and practices. The bodies of some non-human animals are also socially consti- tuted and reflect social inequalities. In order to further examine how social differences are embodied, I want to now consider technological production and reproduction, where debates on power and embodiment are thrown into sharp relief.

Bodies, technologies, gender and nature

The technologising of agricultural space and reproductive manipulation of agricultural animals, and the increasingly common deployment of 'new' reproductive technology in human fertility treatment, has meant that reproduction is an important arena wherein relations between the natural and artifactual are played out.

The bodies of animals and plants have been the subject of human technological interference prior to the advent of modernity in Europe. However, there are particular ways in which capitalism and other formations of social power cast the relations between embodied humans and the environment in which we are embedded. For Marx, labour distinguishes humans from other species and human development took place as a result of a dialectical relation between nature and human abilities to transform the conditions of life through labour. Peter Dickens (1996) expands on these insights through a detailed analysis of the division of labour in contempo- rary capitalism, which structures the ways we work on nature in order to produce the things we need. Organising labour around the production of marketable goods on the basis of increased production and consumption to satisfy the profit motive means that the natural environment is exploited, and nature, including the bodies of plants, animals and even humans, is transformed into objects.

Farmyard tales: The myths of biotechnical breeding

The powers of nature are utilised as production processes, for example, in the making of 'animal products' such as milk, meat and eggs. Biotechnologies are deployed as adjuncts of capitalist process in the cases of reproductive technologies and genetic engineering, for example. Here, the materials incorporated into the production processes are living beings with their own causal powers and properties. In both cases, body parts and elements are separated from their context and manipulated to produce new hybrid entities. Animal 'species and natural being' (Dickens, 1996: 62) is exploited, producing commodities through intervention in the ways in which animals reproduce. The powers of animals (similarly to those of industrialised workers) have been subjected to progressively intensive rationalisation, automation and mechanisation in order to produce commodities for human consumption.

For Dickens (2001), the relationship between capitalism as a social and economic system and physical bodies has never been so close. Capitalism is 'modifying biology in its own image' – bodies of humans, animals and plants are changing physically in significant ways, to satisfy the profit motive. The development of human agriculture can itself be seen as a mechanics of humanising and hybridising nature. However, various animals and plants have undergone more invasive technological manipulation in the past 30 years, as genetic engineering is increasingly the norm in 'enabling' nature to transcend biological limits and become more productive. Dickens (1996: 114–115) argues that capital is increasingly intervening through biotechnology, for example, to regulate the reproduction of agricultural seeds. In the global 'South', traditional praxis has meant that farmers can save seed after harvesting and exchange locally. Bio-diversity prospecting in poorer countries, argues Shiva (1993), means that the new colonialism of biotechnological transnational capital can patent plant life forms and extend commodification into new areas of life. In addition, the manufacture of genetically modified seed ensures increased dependency on transnational capital through patenting and other trade agreements (Shiva, 1998), which displace indigenous practices and knowledge.

Such developments are the latest form of the industrialising of nature and commodification of non-human lifeworlds. As Dickens and Shiva suggest, industrial biotechnologies applied to the bodies of agricultural plants and animals seems set to reinforce or enhance patterns of local and global inequality, rather than contest it, as the myth of the cyborg suggests.

Family fantasies: Some myths of assisted conception

Procreation in humans is also subject to increasing biotechnological intervention. Whilst some contend that new reproductive technologies exemplify the postmodern condition in their destabilisation of scientific certainty (Franklin, 1997: 211), others suggest that the project of science as progressive control of the natural is further exemplified by experiments in conception (Oakley, 2002: 144). What is clear is that there are differential effects across the globe, with the commercialisation of procreation seeming to 'benefit' wealthy Western women, whilst women in the South are often subjected to fertility control (Mies and Shiva, 1993). Historically, there have been technologies applied to limit reproduction of undesirable humans (poor, promiscuous, unmarried, non-white) (Greer, 1984: 279), and this is often still the case.

Some feminists in the 1970s saw great potential in new reproductive technologies. Shulamith Firestone (1988) infamously posited that women must seize control of the means of artificial reproduction in order to liberate themselves. Many were alarmed however at the consequences of applying such technologies to women. The technologies themselves were developed within animal breeding for the meat industry and include artificial insemination by donor, in-vitro fertilisation and embryo transfer (Corea, 1985). Janice Raymond (1993) suggests the 'problem' of Western infertility is constructed through a technoscience discourse that operates to sustain a market for ineffective and potentially dangerous technology. In-vitro fertilisation and embryo transfer involve complex medical procedures (Pfeiffer, 1987: 88), which some consider deconstruct motherhood into a series of passive roles and functions (Greer, 1999). There are risks of anesthesia, surgery, trauma to ovaries and uterus, ectopic pregnancy, unknown effects of hormones, unexpected outcomes such as multiple pregnancies (Price, 1999: 30–48), the increased likelihood of miscarriage, congenital abnormality or caesarian delivery (Edwards 1999: 44), stress and disruption to paid employment (Franklin, 1997: 82). The live birth rate per treatment cycle in the United Kingdom is 18 per cent, and death rates for babies so conceived are 32 per cent higher than those born without IVF (Oakley, 2002: 144).

Whilst the boundaries between animals and technological mechanisms are indeed blurred here, such practices exemplify an extreme faith in science as a modern panacea for 'ills' (Rabinow, 1992). The IVF baby, the woman with the implanted embryo, the genetically modified tomato and the cloned domestic cat are all hybrids, objects of technonatural times. But this does not mean that they spell the end of the problems with diseased and difficult bodies or that they are positive developments, indeed human cyborgs are often undesirable – the product of disease, illness or near fatal injury (Shaviro, 1995). In any case, human times have always been 'technonatural' in that we are dependent on 'nature' for survival and utilise technologies to do so. The iniquitous social systems that shaped the development of modern society, of capitalism, colonialism and gender relations have been reflected in, and are constitutive of, our technological mediation and shaping of 'nature'. Whatever the changes in such technologies of production and reproduction, relations of inequality continue to play themselves out, in and on, the bodies of humans and other animals.

Contesting the posthuman

As we have seen, the social condition of 'becoming cyborg' cannot be separated from the persistent social, economic and political structures of inequality. What I want to address in this final section is the question of our identity, which is an important element of the mythology of posthumanism.

Do we all know that we are 'posthuman' now?

Haraway assumes that cyborg subjects will experience their posthumanism, yet the notion that 'we' become posthuman because we think we are, is deeply unsatisfactory. The posthuman is a highly specific and located phenomenon. The notion of the posthuman as a 'condition' is a universalistic presumption, for any subjectivity of the posthuman affects an incredibly small proportion of the world's population. 'We' are not all post-industrial, and whilst certain communities within first world societies may be enmeshed within information networks, the hybridity of the poor assumes different formations, and these may be distinctly embodied and far from the information super-highway. The posthuman as it is currently usually articulated is, at best, a utopia. However, notions of the posthuman may open up the possibilities of new kinds of dwelling in what Arne Naess (1979) would call 'mixed communities' which include other species.

Hanging out with the non-humans

Before she concentrated on the 'cyborg' as a symbol of the hybridity of socialnature, Haraway tended to characterise nature as a 'witty agent' and drew an analogy with the North American coyote which she saw as a hybrid creature, living on the borders between human society and the 'wild'. Haraway has recently argued that domestic companion animals, particularly dogs, 'more fruitfully inform livable politics and ontologies in current lifeworlds' (2003: 4) than cyborgs or coyotes. What she stresses is that 'companion animals' question the idea of 'society' as exclusively human because they are in relations of 'biosociality' with humans. In my view, any move to a more benign set of relations between humans and domestic animals is premised on a transformation of the institutions and practices of industrial agriculture in which the majority of the domestic animal population is embedded. However, Haraway and others (Franklin, 1999) are right to suggest that human relations with companion animals enable us to rethink the boundaries of our 'humanness' in positive ways.

Comparisons between computation and the animate lifeworld can be made, but this does not mean artificial and human/animal life are the same. Our human minds are embodied, and cognitive beings, like humans and dogs, are alike in their embodied realisation and actualisation of self(hood). The boundary to the integration of humans with machines is the distinct difference of their physical presence – because humans and animals like dogs are embodied, they think differently (see Varela, Thomson and Rosch, 1991: 98–99).

Questioning human centredness has much to recommend it, but we cannot simply dissolve the categorisation of human and non-human that is an edifice of political power, economic organisation and social stratification. We need to examine institutions of social power and account for persistent differences and inequalities in any attempt to realise an embodied and ecologically embedded posthuman future.

10. Myth and HIV Medical Technologies: Perspectives from the 'Transitions in HIV' Project

MARK DAVIS AND PAUL FLOWERS

Introduction

This chapter uses the advent of Highly Active Anti-Retroviral Treatment (HAART)[1] to explore myth and biotechnology. With reference to qualitative interviews with people using HIV (Human Immuno-Deficiency Virus) treatment, we will address how myth is implicated in imagining what can be achieved through biotechnologies particularly in connection with hopes and expectations about the life course.

There have been significant achievements in the medical management of HIV since HAART was introduced in the mid- to late 1990s, at least in countries that can afford it. For example, in the United Kingdom, deaths due to AIDs (Acquired Immune Deficiency Syndrome) have declined from 1531 in 1994 to 266 in 2003 (Health-Protection-Agency, 2004). Treatments have brought about the much-hoped-for end of the AIDS crisis in the affluent, global North. Because of these treatment effects, the epidemic has been figured as a subsiding crisis and normalised (Rosenbrock et al., 2000). Researchers have used terms such as 'living after crisis' (Rofes, 1998), 'post-crisis' (Race, 2001), 'post-AIDS' (Dowsett and McInnes, 1996) as ways of understanding these changes. For some, HIV treatment is the realisation of cure, albeit partial, and therefore a fulfilment of the promise of biomedicine. For example, analysts have argued that the clinical diagnostic term, AIDS, is no longer relevant (Greene and Ward, 2002).

However, hopes that HIV treatment is a solution to the epidemic of HIV may have mythical qualities. Despite effective treatment, in 2004 alone 3.1 million people died of AIDS (39.4 million people are estimated to be living with HIV worldwide (UNAIDS/WHO, 2004). A recent review of the psychosocial aspects of living with HIV concluded that despite significant progress in HIV treatment, people with HIV and their carers faced challenges of uncertainties owing to the evolving and sometimes ineffective technology (Green and Smith, 2004). The post-treatment situation of the epidemic is therefore a complex mixture of an improved capacity to treat HIV and problems of treatment access and long-term use. The effects of HIV treatment technology are circumscribed by social structure and the uncertainties and limitations of the technology itself. This perspective resonates with the point made by Mosco in connection with the magical qualities often ascribed to computer-related technologies such as the Internet. The question of techno-mythology has also been implied in commentary on the social aspects of the HIV and AIDS epidemic. Susan Sontag's book *AIDS and Its Metaphors* mapped out the connections between HIV and social responses to the threat of plague (1988). In the logic of cultural responses to contagion, the danger of plague is seen to arise from connections with foreign places. Disease is therefore always attributed to the foreign other. Paula Treichler, interested in homophobia in biomedical discourse about the HIV epidemic, pointed out how HIV and AIDS was revealed through a binary logic of othering, for example, 'vice and virtue'; 'us and them'; 'love and death'; 'certainty and uncertainty' (1999: 35). In particular, Treichler enumerated the various mythical explanations for the cause of the HIV epidemic. At one time or another, these have included 'a CIA plot to destroy subversives'; 'a soviet plot to destroy capitalists'; 'a golden opportunity for science and medicine' (12–13). Cathy Waldby has addressed the presumptions and biases of HIV biomedicine 1996). Key to her argument is how myths about the body of the notional heterosexual man are furthered in biomedical practices of constructing the supposedly vulnerable, permeable and inferior female and homosexual bodies.

'Transitions in HIV management'

This chapter extends the argument about myth and HIV treatment by drawing on the published work from research funded by the ESRC 'Innovative Health Technologies Programme' (Ward, Davis and Flowers, 2006) (see: http://www.york.ac.uk/res/iht/). Called the 'Transitions in HIV Management: The Role of Innovative Health Technologies', the research comprised 68 interviews with people with HIV living in Glasgow and London (including gay men, injecting drug users and people from Africa residing in the United Kingdom); 16 interviews with HIV clinicians and treatment advocates; a literature review of the ways in which risk identities are operationalised in medical science underpinning HIV treatment innovation; and a textual analysis of the social marketing of HIV treatment. One of the central aims of the research was to provide an account of the impact of effective HIV treatment on HIV positive people's experience of the infection, their changed understanding of their bodies in relation to HIV medical technologies and the reconfiguration of HIV and AIDS identities.

The findings from the research have been summarised in three publications concerning the accounts of HIV treatment prescribers (Rosengarten, Race and Kippax, 2004); perspectives on qualitative interviews with people from Africa residing in the United Kingdom (Flowers et al., 2006); and the perspectives of gay men with HIV living in Glasgow and London (Davis, Frankis and Flowers, 2006). Doctors prescribing HIV treatment have identified difficulties, including the uncertain qualities of treatment, managing side effects[2] and drug resistance in the long term. People from Africa with HIV point out that HIV treatment is one among several pressing issues, which include parenting, residency status and fear of prejudice. For some interviewees who had experienced HIV/AIDS in Africa and for those who were not allowed to continue to stay in the United Kingdom, HIV diagnosis meant likely death. The technical solution to HIV infection is therefore circumscribed by citizenship raising a contradiction in post-HAART discourse linked with residency status. Interviews with gay men with HIV reveal that the experience of HIV treatment is heterogeneous. For some, disappointment with regard to treatment effects raises psychosocial concerns to do with uncertain health and the return of the prospect of illness and death.

The following sections consider three themes for the 'Transitions' study to explore aspects of the post-HAART situation for people with HIV. The first section discusses some other challenges to post-HAART discourse to know mythical constructions of a biotechnical solution to HIV infection. The next section considers the connections between life expectations and the uncertainties of HIV treatment. The last section raises some further points about the intersection of post-HAART discourse and constraints on treatment access.

Challenging post-HAART myths

Several researchers have discussed the social aspects of the introduction of HAART and the related shifts in life expectations and AIDS identity. These studies suggest that the idea of HIV treatment as cure has a mythical quality. A study from France explored the personal accounts of people with HIV diagnosed before and after the 1997 watershed (Pierret, 2001). Pierret interviewed a group of long-term survivors in 1996 and 1997 and compared these accounts with those of another group interviewed in the early 1990s (2001). The group from the early 1990s constructed their narratives in ways that helped them to cope with the uncertainties of the life course with HIV/AIDS. The long-time survivor group interviewed in 1996/1997 constructed uncertainty as an aspect of the past. This difference was taken to signify a change in personal engagements with the question of being-in-the-world or security for people with HIV infection. Pierret attributed these substantive differences to '...confidence in their stable state of health' among the long-term survivors and possibly also to the advent of treatment in 1996 (177). In 1997/98 Trainor and Ezer conducted a study of people who thought they were going to die, to explore the treatment-related turn around in life expectations arising out of effective treatment (2000). Unlike Pierret, Trainor and Ezer found that the removal of the prospect of death created uncertainty for the interviewees. This counter-intuitive aspect of turnaround was understood as arising for interviewees because death had been a given. Rejuvenation of life expectations was upsetting because it undermined the accepted biography of someone with AIDS. In the United Kingdom, Flowers has discussed this counter-intuitive aspect of turnaround in terms of a shift from 'death sentence' to 'life-sentence' with treatment, underlining the ongoing and increasing challenge of living with HIV (2001). Davis has discussed using HAART with reference to adherence with treatment regimes, the medical technologies used to monitor treatment effectiveness, and the related engagements with systems of expert knowledge (Davis, 2007). A review conducted on behalf of the British HIV Association (which advises HIV doctors on prescribing practice), suggested that a key psychosocial challenge for people with HIV was the prevailing sense of uncertainty associated with treatment (Green and Smith, 2004). Considering identity and the advent of HAART, Watney has argued that constructions such as post-AIDS and post-HAART oversimplify lived experience (2000). In particular, the notion of a technological watershed in the history of the HIV epidemic both limits yet maintains counter-watershed constructions of living post-HAART and obliterates the basis for an interrogation of techno-uncertainties.

Resistance to the idea of treatment = cure was evident in the interviews conducted for the 'Transitions' project. For example, treatment was welcomed but interviewees challenged the idea of a straightforward post-HAART situation for people with HIV:

> To call it a treatment is a real misnomer because what happens is the press gets hold of it and they start writing articles about new treatments and they're not treatments at all because treatment I think infers that it's a cure [and that] HIV has now become a manageable thing. HIV therapies aren't either of those things. At the moment it's a short stop thing that in most cases has awful side-effects, some in the short-term and some in the long-term that people don't even know yet...People were saying these are fine. You might get diarrhoea. A bit of nausea or whatever or something, but they're all manageable, there's nothing life-threatening and of course it's not the case. The fact is that kidneys have got

holes in them and leak because of DDI. I was prescribed DDI and no-one knew such a thing would happen...I don't think either it's a treatment or it's a management of HIV. A short stop-gap to try and give you a bit of extra time in the hope that newer and better drugs will come along that will be able to treat it or manage it long-term or maybe have a cure. (Davis, Frankis and Flowers, 2006)

In this example, the interviewee makes a separation between their own experience and popular constructions of HAART and its effects. On that basis, HAART falls short of the status of 'treatment'. Instead, HAART is a way of managing the body with HIV in time, until a more effective form of biotechnical intervention can be found.

In contrast with the globalising idea of treatment = cure, the HAART experience concerns the unfolding quality of technological innovation that underpins HAART. In the next example, the interviewee identifies the management of technology in time as the method of treating HIV:

my actual medical care the whole way through, I mean the seventeen, twenty years, that I've been HIV has been exceptionally good in terms of my medical care. And by that I mean whenever there has been something wrong with me, things have moved quickly. Whatever treatments were available I had them...as the technology has developed I had access to them. (Davis, Frankis and Flowers, 2006)

In this account, technological developments have made biography possible. Survival is attributed to fortuitous technical innovation. This account sits in contrast with the idea of the advent of treatment as a historical moment, 'turnaround' narrative and the investment of these ideas in the promise that biomedicine will deliver a cure. As Mosco suggests, myths can both reveal and conceal (2005). They can help us to imagine the potential of technologies, but they can also obscure power relations. In connection with HIV treatment, the myth of cure implies a treatment with the power to combat HIV infection. The examples discussed above suggest that this notion of treatment is only partly relevant for people living with HIV. People using treatment do not assume cure. Rather, they focus on the reflexive management of innovating HIV treatment, a practice that embraces but exceeds the treatment = cure myth.

Other research has suggested that a refusal of the promise of biotechnology is a superior way of using HIV treatment. Deploying the idea of narrative, Ezzy has analysed the personal accounts of people living with HIV in Australia (2000). He suggested that interviewees adopted one of three narratives for dealing with the advent of HAART and its attendant uncertainties: 'linear restitution'; 'linear chaos'; and 'polyphonic'. Each of these narrative approaches to uncertainty implies different engagements with biomedical technology. Linear restitution narrative constructs HAART as restoring the life prospects of the person with HIV. The linear chaos narrative is held to reflect uncertainties intruding into and therefore undermining life expectations. Ezzy proposed that both of the linear narrative approaches reflect modernism, or the hope of a gradually improving technology, either realised or undermined. Linearity in technological improvements is therefore a necessary myth for these kinds of stories. Polyphonic narrative is quite different. It is held to reflect an acceptance of uncertainty and suspends the idea of a linear biography for a focus on the 'here and now'. By contrast, linearity is too fragile and open to disappointment. Ezzy therefore proposes that polyphony, and therefore a kind of post-modern reflexivity, is a more 'robust' form of engagement with HAART-related uncertainties. But Ezzy is also implying that after myth there is no biography. Letting go of hope as an organising principle of biography implies a form of life without an imagined future. Such reflexive practices of uncertainty raise questions about lived experience in the ambivalent conditions of technological innovation. Unlike Ezzy's polyphonic narrative, interviewees for the 'Transitions' research seem invested in the promise of a gradually developing biotechnological solution to HIV infection.

Expectations and uncertainties

The emerging sociology of expectations literature has begun to map out the social and cultural aspects of innovating technology (Brown and Michael, 2003; Webster, 2002). Through interviews with researchers and clinicians involved in xenotransplantation, Brown and Michaels discuss how 'future promise' and 'techno-determinism' underpin how innovation plays out in the arena of biotechnology. In Mosco's terms, expectations of a biotechnical fix have the proportions of a myth. Brown and Michaels advocate a critique of biotechnological innovation that engages with uncertainty and how future expectations are metaphorised and colonised.

However, the sociology of expectations literature has some limitations in relation to the present case of HIV treatment. The approach elides macro-analyses of expectations of evolving technology with personal biography, which somewhat obscures the lived experience of those whose life course is dependent on technology. For example, in relation to the treatment = cure myth, interviewees in the 'Transitions in HIV management' project talked about their personal strategies for making treatment work for the purposes of survival. They also indicate the methods of psychological self-care connected with the uncertainties entailed in using HAART. In particular, it seems that for some people with HIV in this situation, life expectations are less relevant:

> I have to turn my life around really because one thing it does do which is I think very sad is you don't make any plans for the future. No long term plans. Now you know, even at my age I'm sure if there is nothing wrong I would still want to buy a house, get a mortgage, do a bit of work over, do you know what I am saying but there is nothing, you are just existing. (Davis, Frankis and Flowers, 2006)

Much like the polyphonic narrative identified by Ezzy, this interviewee has set aside the idea of planning a future for a focus on living now (2000). Living with HIV treatment is therefore a challenge to sense of self. Writing in the period prior to effective treatment, Roth and Nelson characterised AIDS as undermining the basis of social existence and identity in addition to causing physical harm (1997: 161). In this regard, they refer to 'ontological health' which is taken to mean the sense of personal security for people living with HIV. This concept of ontological health connects with ideas concerning reflexive modernisation discussed by Giddens. In particular, Giddens uses the concept '...ontological security' to draw attention to the personal challenges of trust and security that individuals face in rapidly technologising and globalising societies (1990: 92). Specifically, ontological health resonates with ideas concerning how the individual has to engage with challenges to the '...continuity of their self identity and in the constancy of the surrounding social and material environments of action' (92). In this regard, Giddens is arguing that technological innovation has a dual effect because it is desirable or even necessary but that it also provides the conditions for increased uncertainty and, therefore, anxiety. As Beck and Beck-Gernsheim explain in connection with genetic technologies, 'As the possibility of genetic prediction grows, so too, paradoxically, does biographical uncertainty' (2002: 139). In this regard, HIV treatment foregrounds the psychosocial challenge of (in)security.

However, it is also the case that this sense of future interrupted is not restricted to the post-HAART situation. For example, ontological insecurity related to biotechnology was an aspect of HIV prior to the advent of treatment. By the late 1980s, a diagnostic/prognostic system had developed that placed people with HIV into categories such as 'seroconversion'; 'latency'; 'asymptomatic'; 'AIDS' (Macintyre, 1999). Each of these categories has since altered as diagnostic knowledge and techniques have developed. By the post-HAART era, some argued for the eradication of labels like AIDS (Greene and Ward, 2002). But these categories and their refashioning have important implications for personal security. For example, writing about the period prior to HAART, Macintyre argued that as medical technologies and scientific practices have evolved in the management of HIV infection, some patients found their hopes undermined. When HIV and AIDS were less well understood, long-term survivors (people known to have been infected with HIV but have not become AIDS patients yet) were a focus in clinical research. Such patients (and their clinicians) hoped that somehow they were able to defeat the virus in their bodies. However, as mortality data accumulated and medical knowledge expanded, prognoses were gradually lengthened, erasing the label 'long-term survivor' and providing a nearly universal prognosis of eventual death and therefore undercutting hopes of survival. In this situation of developing but disappointing knowledge, hopes for effective treatment took on extra force. In addition, the uncertainties derived from the probabilistic methods used by science were also a source of distress for patients.

The connection between medical technology, hope and conduct has therefore always been a feature of HIV and AIDS. No less in the post-HAART situation, treatment technologies mingle with biographical time. For example, people who find they have HIV infection have blood tests at regular inter-vals to help their clinicians assess when they may need to begin HAART. Two main tests are used for this purpose: a viral load blood test[3] and a CD4 blood test (AIDSmap, 2003). Viral load tests measure the amount of virus in the blood, indicating the amount of viral activity. CD4 blood tests measure a part of the immune system to provide an indicator of decline in health. Once treatment has begun, patients are required to have viral load and CD4 blood tests to monitor treatment effectiveness. Genotyping can also be included to monitor the development of drug resistant forms of HIV.[4] HAART and its related technologies therefore provide the basis for an intensification of the formularisation or measurement of hope. As pointed out by others, these blood tests therefore provide new possibilities for the self-regulation of the patient with HIV, connecting biographical practices with not just the advent of HAART, but the ongoing use of such medical technologies (Race, 2001).

Interviewees in the 'Transitions' project discussed the uncertainties of treatment and implications for self-care. In the next example, the interviewee reckons that a cultivation of the present is a method for existing with uncertainty:

> Half the battle I think is in your mind too. I mean it's a terminal disease. It's gonna get me eventually... but you can't linger on it. I feel good today. You're gonna live good today. (Davis, Frankis and Flowers, 2006)

A traditional psychological interpretation of such an approach might be denial. Such denial would be seen as a defence mechanism that helps the harried ego to cope with the potentially overwhelming anxiety of possible treatment problems. This example is also fatalistic in the sense that it speaks of the uncontrollable dangers that might arise in the future. However, there is a subtle kind of moralising in play in interpretations that turn to denial and fatalism. Other researchers have identified how a moralism has invaded the discursive regulation of citizens with chronic illness (Galvin, 2002). Health subjects who fail to take action to protect and enhance their health are seen as individually responsible for adverse health outcomes. Central to this discourse is the model health citizen, thoughtful and vigilant, acting on themselves to optimise their bodies and minds for the personal project of health and well-being. In connection with HIV, the treatment = cure myth fuses a sense of moral conduct of the self, the emotional life of the individual and personal hopes and expectations for the life course. In contrast to the ideas of denial or moralising discourse, the example from the 'Transitions' project appears to claim a freedom to exist with the limitations and uncertainties of HIV treatment and therefore biotechnology in general.

The post-HAART idea of a curative biotechnology intensifies the moralising aspects of biomedical discourse. In particular, epidemiologists have long recognised that epidemics have a 'natural' history where they are limited by the death of the host. HAART, however, means that there is a growing population of people with HIV, leading epidemiologists to speculate that there is also an increasing risk of the spread of HIV (Boily, Godin, Hogben, Sheer and Bastos, 2005). In a related sense, researchers have become worried that because of the satisfaction of hopes concerning the cure of AIDS, people at risk of and with HIV might stop using condoms during sex, therefore also leading to an increased spread of HIV. This chimera of hope and post-HAART risk production is called treatment optimism (International-Collaboration-on-HIV-Optimism, 2003). It typically relies on forced choice surveys to measure treatment optimism and its association with reported risk behaviour, although qualitative research has also been conducted (Davis, Frankis and Flowers, 2002; Rosengarten, Race and Kippax, 2001). The idea of treatment optimism is a kind of techno-myth in Mosco's sense. It springs from the hope invested in an effective biomedical solution to the HIV epidemic and resonates with a more general notion of the imperilled promise of biomedical technology. Treatment optimism establishes how, in the hearts and minds of errant citizens, the benefits of medical technology can unravel itself. Treatment optimism therefore reveals a dystopian/utopian paradox that shadows biotechnical innovation and perhaps technology in general. But importantly, the post-HAART situation focuses on individual citizens and how this dual quality of technology is realised through their agency.

However, the accounts of engagements with uncertainty and expectations derived from the 'Transitions' project reveal a different moral subjectivity informed by the uncertainties of HIV treatment technology. These interviewees resist the idea of a definite biotechnical watershed in the history of HIV or their own biographies. In place of a mythical, absolute cure and the associated moral imperatives on conduct, these interviewees are more interested in managing the self in conditions of uncertainty and evolving technology.

Citizenship

It also seems that the myth of a biotechnical solution to the HIV epidemic is undercut by social factors. Living in the United Kingdom does not necessarily guarantee access to HAART. In the next example, the interviewee reflects on what happened when a dispute arose over their status as an asylum seeker:

> So I went there and met my counsellor. She was so nice to me, I didn't have any problems. Now here comes when the time for me to get the treatment because my CD4 count was so low – it was almost 50, but I was looking fit. The first doctor I saw, she was a lady, said to me, what about your immigration status and I said, I don't have anything, I was just an illegal immigrant. We had an argument and she said, no we can't give you treatment. (Flowers et al., 2006)

This example reinforces the idea that the capacity to believe, take refuge in, and enjoy the biotechnical mythology associated with HAART is patterned by citizenship. HIV is a global epidemic, increasingly superimposed on social exclusion (Green and Smith, 2004). It is difficult to gauge the size of this form of exclusion from qualitative research. For example, such people may be unwilling to volunteer for research about HIV treatment. However, claims about HAART in particular and biomedical innovation in general need to be tempered with an awareness of its structural constraints.

In addition, people from Africa living in the United Kingdom found the promise of HAART did not deflect the symbolic status of HIV infection as a death sentence. In this example, the interviewee reflects on the loss of hope for the future associated with HIV:

> I think for me, I just felt my life was gone, as a youth, as a young woman aspiring for so much in life. I thought that was the end of it. At that time, when I was diagnosed, I wasn't the type who was into make-up at all. That day, I bought... I used to buy make-up and keep it in my drawers. I got home and all night that's what I was doing...putting all the foundation and doing my face up! I felt I was denied life. I felt I'd never have a child, never have a family, I'd never get married, and I'd never fulfil my career or education prospects. My life was gone, and the way it was handled, the stigma around at the time, the death threats – coming from Africa, I'd seen so many people dying that to me that was the end of my life actually. (Flowers et al., 2006)

For this interviewee, the post-HAART engagement with the normalisation of HIV as a chronic, manageable illness is not relevant. HIV infection is understood in light of the experience of living in Africa. In addition, uncertain residency status could undercut the benefits of HAART dramatically. In this quotation, the interviewee reflects on the distress concerning not knowing for how long they would be able to stay in the United Kingdom and therefore how long they could expect to use HAART:

> Well you know it is like: all those problems you never knew you had, they have just now come. Me, I wish I wish I'd never got tested, you believe me, I would have maybe I would have gone and get sick and die without knowing it's the one that is killing me. Because you know ahh...it's not worth living. It's not worth living, honestly; as much as like now I am on medication all the time, I am an asylum seeker here. (Flowers et al., 2006)

This example underlines how residency status prefigures engagements with HIV diagnosis. It seems that residency status also contextualises the use of HAART itself:

> Although I had, at some point in May, my viral load was undetectable, because of the difficulty of the side affect of...this medication although sometime you just keep [you miss it] yeah sometimes this side affects, they're there pain in the stomach and neuropathy and so on, and few time I was missing the doses and then my viral load started going up again, up again...and I started talking to the, specialist clinic, and I said I cannot continue with this thing it is giving me a lot of trouble, mm, some kind of force is keeping because of the fear of the side affects and the problem I get from the side effect., huh and they say well continue, we'll monitor, we'll monitor the situation, because I was doing, well when I was in hospital I was doing well, but you see it's difficult when you get out of the hospital, in hospital they monitor your medication very strictly and then when you are out on your own, like I live alone, it's difficult. Sometime you cannot cook anything, you are supposed to take with food. (Flowers, Rosengarten, Davis, Hart and Imrie, 2004)

The post-HAART HIV epidemic is therefore fractured by social structure, amplifying the mythical qualities of the idea that treatment = cure. It is estimated that about 5.8 million people in the world need HAART (UNAIDS and WHO, 2004). However, access to HAART is compromised by its cost and the challenge of providing it in different health-care settings throughout the world. In 2003, UNAIDS set in motion a global campaign to increase treatment access. Called *3 x 5: Make It Happen*, the campaign aims to increase the number of people on treatment to 3 million by 2005 (http://www.who.int/3by5/en/). By June 2005, it was estimated that 1 million people were using HAART in resource-poor settings, up from 440,000 in 2003, but short of the campaign target (UNAIDS and WHO, 2004, 2005). In transitional countries, treatment-related uncertainty has to do with emerging but fragile economic systems, where expensive HIV treatments can be funded only periodically. For example, until recently in Serbia, patients could not expect to have constant access to HAART (Rhodes, Bernays, Davis and Green, 2006). For some months in a year, there were restrictions in funding for health care and therefore cessation of HAART. The particular structural situations of countries in economic transition therefore influence access to HAART, as with other aspects of public health responses to the HIV epidemic (Rhodes and Simic, 2005). Considering some of the global-scale challenges of treating people with HIV makes the myth of treatment = cure seem pale and flimsy.

Conclusions

HAART represents the partial realisation of hopes for a biotechnical solution to the HIV epidemic. But as Mosco suggests, the 'taken for granted' qualities of biotechnical discourse attached to the advent of HAART and its effects is both revealing and concealing. The idea of a biomedical solution to the HIV epidemic is necessary for the self-care practices of people with HIV. But such hopes are made provisional because of side effects and technical uncertainties, creating new psychosocial dilemmas for the self with HIV. In some situations, uncertainties require people with HIV to let go of the myth of techno-optimism, erasing the prospect of future biographical time. Such post-myth deletion of biography suggests a more general effect of the uncertainties that come with technological innovations in the area of health. Conversely, the myth of techno-optimism informs the idea that through the satisfaction of hopes connected to HAART, people run amok, failing to have safer sex and safer injecting drug use, spreading HIV and therefore undermining the biotechnical ascendancy in the management of HIV. In addition, treating HIV is undermined by citizenship, the economics of drugs development and unequal global access. A simple notion of biotechnology triumphant masks what is going on for people with HIV.

Notes

1. Highly Active Anti Retroviral Treatment. HAART is comprised of combinations of different drugs that act on HIV infection in different ways but together, reducing viral activity and allowing the immune system of the person with HIV to recover and resist infections and cancers. In practice there are many different combinations used depending on the patient's tolerance and drug-resistance (BHIVA, 2005).

2. Like many treatments. HAART has side effects, some of which are severe. For example, diarrhoea, headaches, nightmares, changes in body appearance, heart disease (NAM, 2003).
3. One feature of HIV medical technology is regular blood testing to monitor viral activity and therefore the effectiveness of HAART in the body (NAM, 2002b). Viral load is an index of viral particles in blood used to infer the amount of HIV in the body. 'Undetectable viral load' is the recommended treatment target (BHIVA, 2005).
4. HIV is a retrovirus which means that it has the property of mutating, sometimes into forms that resist the effects of HAART (Bangsberg et al., 2003).

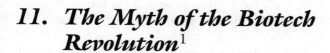

11. The Myth of the Biotech Revolution[1]

PAUL NIGHTINGALE AND PAUL MARTIN

Introduction

The existence of a medical 'biotech revolution' has been widely accepted and promoted by academics, consultants, industry and government. It has generated expectations about significant improvements in the drug discovery process, health care and economic development that underpin a considerable amount of policy-making. This chapter presents empirical evidence, from a variety of indicators, which shows that a range of outputs have failed to keep pace with increased R&D spending and rather than producing revolutionary changes, medicinal biotechnology is instead following a well-established pattern of slow and incremental technology diffusion. Consequently, many expectations are wildly optimistic and over-estimate the speed and extent of change, suggesting the assumptions underpinning much contemporary policy-making need to be rethought.

The biotech revolution's role in policy-making

Over the past decade consultants, policy-makers, academics and industrialists have promoted a model of technical change in which biotechnology in general and genomics in particular are revolutionising drug discovery and development (BIGT, 2003; Commission of European Communities, 2002; OECD, 1997, 1998, 2004). This 'revolutionary' model has generated widespread expectations that biotechnology has the potential to create an increased number of more effective drugs and bring about radical changes in health care, involving a shift from reactive to preventative and more personalised medicine (Bell, 1998, 2003; Collins, 1998; Department of

Health, 2003; Lenaghan, 1998; Lindpaintner, 2002b; Milburn 2001). This, in turn, is expected to stimulate a shift in the industrial structure of the pharmaceutical industry from large drug companies to networks of bio-technology firms agglomerated in regional clusters (DTI, 1999; Enriquez and Goldberg, 2000; Tollman, Guy, Altshuler, Flanagan and Steiner, 2001). Together these changes are expected to lead to improved health and wealth creation (BIGT, 2003; DTI, 1999, 2001; House of Commons Science and Technology Committee, 1995; Tollman, Guy, Altshuler, Flanagan and Steiner, 2001).

These high expectations now underpin much science and technology policy at the OECD (1998, 2004), in the United States (Collins, 1998), the EU (Commission of the European Communities, 2002; DTI, 1999), and develop-ing countries. Agencies at the regional, national and supra-national levels are investing heavily in biotechnology and genomics in order to establish a foot-hold in what is seen as a key part of the New Economy (Dohse, 2000; DTI, 2001; Giesecke, 2000; Senker, Enzing, Joly and Reiss, 2000) Policy takes a number of forms, including dedicated research funding programmes, fostering knowledge/technology transfer, financial and technical support for start-up firms and regional clusters, R&D tax credits and lower regulatory hurdles (ibid.). In the United Kingdom, the recent report by the Bioscience Innovation and Growth Team (BIGT) argued that in order to realise the great potential of the molecular biosciences, there is a need for significant changes in the relationship between the National Health Service and industry, to allow easier clinical trials, and earlier and cheaper access to new medicines (2003). Similarly, the idea of a biotech revolution has increased policy emphasis on closer networking between university researchers and industry, and focused funding on research that can be directly applied (DTI, 2000).

In this chapter we argue that the 'biotech revolution' model of technical change is unsupported by the empirical evidence. Instead, biotechnology is following a well-established historical pattern of slow and incremental technology diffusion. In making this case, we are not denying that there has been a substantial change in the biological sciences and the organisation of R&D within industry. This is obviously happening. However, the translation of this science into new technology is far more difficult, costly and time-consuming than many policy-makers believe (see also Horrobin, 2001, 2003).

Changes in the pharmaceutical innovation process: The evidence

It is possible to measure the impact of biotechnology using indicators of scientific and technological activity such as patents, scientific publications

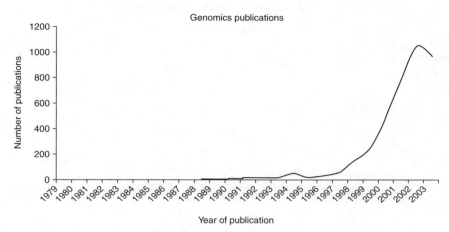

Figure 11. 1. Changes in the number of scientific publications in genomics.

Source: BIOSIS.

and drug launches. As part of a study funded by the UK Economic and Social Research Council (ESRC), we have analysed a range of these indicators along the pharmaceutical innovation process. They show that as one moves along the innovation path from basic research to target discovery, target validation and into clinical development, evidence for a biotechnology revolution rapidly diminishes. A sample of our data, generated by Surya Mahdi (2004), is shown below:

Figure 11.1 shows a sample of our data, generated by Surya Mahdi. This shows the substantial increase in bioscience publications associated with genomics. This clearly indicates a major, and possibly revolutionary, change in *some* of the scientific inputs to drug discovery.

However, when we move further along the innovation path we observe slow and incremental change. Figure 11.2 shows the number of therapeutically active compounds patented each year between 1978 and 1998. The bottom line shows the data for the US Patent Office (USPTO) classes 424 and 514 in the period 1978–2002. This is the main patent classification for therapeutically active compounds, and is used as an indicator of the number of small molecule compounds considered attractive enough to warrant patent protection, but not necessarily viable enough to enter development.

Although we can see a steady rise in the number of patented compounds, this increase in output needs to be interpreted with care, as it does not take into account substantial increase in research spending or changes in the regulatory environment. Figure 11.3 shows that during the same period as the approximate seven-fold increase in patenting, R&D spending increased roughly 10 fold. Even if we take into consideration the expected lag of

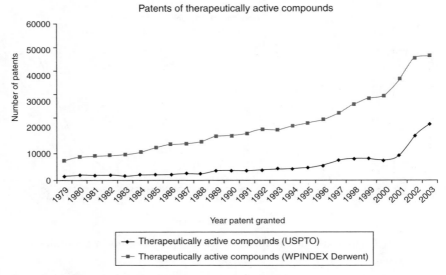

Figure 11.2. Changes in the number of patents of therapeutically active compounds.

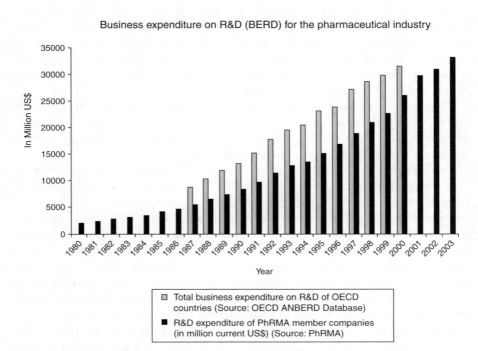

Figure 11.3. Increases in research and development (R&D) expenditure.

Source: OECD and PhRMA.

between four and eight years between R&D investments and patenting in these USPTO classes, there is no evidence of dramatic improvement. On the contrary, we find a decline in R&D productivity as measured by the number of patents per dollar of R&D expenditure. Assuming a relatively constant relation between research and development spending, this indicates a possible decline in research productivity, at least in the short term.

This finding needs to be analysed very carefully. Previous historical studies of major technical changes highlight how new technologies typically produce fast, but localised, quantitative improvements in productivity that are highly visible (Rosenberg, 1979). These are then followed by slow, but often more substantial, qualitative changes that are much more difficult to detect (ibid.). Early applications of biotechnology involved the pharmaceutical industry picking the 'low hanging fruit' and these new technologies are now being used to work on much more complex biological problems that were previously too difficult for R&D to address. What the data suggests, therefore, is that any *qualitative* productivity increases that biotechnology has brought to R&D have not kept pace with the increased complexity of the problems that the pharmaceutical industry and its regulators are now addressing, producing a quantitative decline rather than a revolution.

If we go further on and look at the number of drugs that were actually approved by the FDA in the period 1983–2003, as shown in Figure 11.4, we

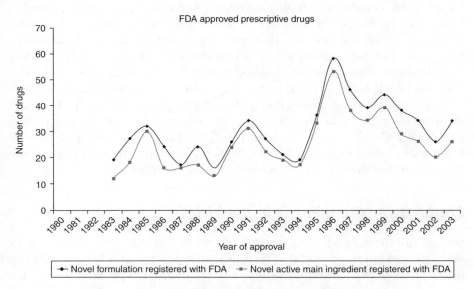

Figure 11.4. Number of FDA approved prescription drugs 1982–2003.

Source: FDA.

see an increase until the mid-1990s, followed by a sharp decline, so that roughly the same number of drugs were approved in 2002 as two decades earlier. When this is set against the substantial increase in R&D expenditure that took place between 1970 and 1992 (i.e., allowing for the 8–12 year lag between research investment and new product launches) there is further evidence of a decrease in productivity, rather than the revolutionary increase we have been told to expect. The peak in the mid-1990s needs to be interpreted within the context of shifting regulatory goal posts, following the Prescription Drug User Free Act (1992) and the FDA Modernisation Act (1997) which allowed accelerated approval and fast track registration. This might be expected to produce a short-term increase in approvals, but such fine-grained analysis is well beyond the limits of this data.

Finally, it is worth examining the number of successful novel biopharmaceuticals that have reached the market since 1980, as these are some of the most tangible fruits of biotechnology. Table 11.1 gives details of therapeutic proteins and antibodies that sold more than $500m a year in 2002/2003 and shows that only 12 recombinant therapeutic proteins and three monoclonal antibodies have become widely used since 1980. Moreover, it is worth noting that three of the therapeutic proteins were already characterised biologicals in 1980 (marked *), with biotechnology simply leading to new production techniques. In other words, the widespread diffusion of recombinant DNA in the 1980s only resulted in a handful of successful new biological drugs. The pattern with MAbs suggests that it can be nearly 25 years before a key scientific innovation becomes effectively translated into new therapies, again suggesting that benefits are hard won. The limited impact of bio-pharmaceuticals on health care has recently been highlighted by Arundel and Mintzes (2004) using data from *Prescrire*, which unlike FDA data evaluates the performance of new drugs *relative to pre-existing treatments*. This data suggests that despite huge investments only 12 biopharmaceuticals evaluated between January 1986 and April 2004 were better than 'minimal improvements' over pre-existing treatments. Taken together this empirical evidence provides no support for the notion that there has been a biotechnology revolution.

Understanding what is happening

Several important questions emerge from this analysis. First, why have so many people got their model of technical change wrong? A key factor is the need for innovators and their sponsors to create high expectations in order to get access to the very considerable resources (money, people, intellectual property) required to develop new medical technologies. No one is going

Table 11.1. Therapeutic proteins and monoclonal antibodies with sales of more than \$500 million in 2002/2003

Product	First launched by	Annual sales 2002/3 (\$m)	Launch date
Recombinant therapeutic proteins			
*Recombinant human Insulin	Lilly	5340	1982 (US)
*Recombinant human growth hormone	Genentech	1760	1985 (US)
Interferon α	Roche and Schering-Plough	2700	1986 (US)
Erythropoeitin	Amgen/Johnson and Johnson	8880	1989 (US)
Granulocyte-colony stimulating factor	Amgen	2520	1991 (US and EU)
Blood Factor VIII	Bayer	670	1992 (US)
Interferon β	Berelex (Schering AG)	2200	1993 (US)
Glucocerebrosidase	Genzyme	740	1994 (US)
Follicle stimulating hormone	Serono and Organon	1000	1995 (EU)
Blood Factor VIIa	Novo Nordisk	630	1996 (EU)
TNF receptor binding protein	Amgen	800	1998 (US)
Lutenising hormone	Serono	590	2000 (EU)
Monoclonal antibodies			
Rituximab	Genentech/IDEC	1490	1997 (US)
Infliximab	Centocor	1730	1998 (US)
Palivizumab	MedImmune	850	1998 (US)

* Already characterised biologicals in 1980.

to invest in a start-up company, or a large-scale scientific endeavour, such as the Human Genome Project, unless they genuinely believe it has the potential to yield significant returns in a defined timescale. The emergence of the biotechnology industry has rested heavily on the creation of these high hopes and many people in the sector have been active in promoting the idea of a biotech revolution. Management consultants, financial analysts and venture capitalists all clearly have a vested interest in hyping new technologies. Similarly, the promise of a biotechnology revolution provides government policy-makers with simple, but as our analysis suggests, probably ineffective ways of promoting regional development, improved health care delivery and economic growth. The failure of social scientists is less excusable.

Having said this, it is important to note that not everyone has believed all the hype surrounding biotechnology and genomics (e.g., Horrobin, 2001, 2003; Nature Editorial, 2003; Williams, 2003). Within the pharmaceutical industry, the debate has been far more nuanced and many people have pointed out the mismatch between expectations and reality, and have stressed the very long and difficult processes involved in bringing new drugs to market (Drews, 2003; Kubinyi, 2003; Lindpaintner, 2002b; Nature Editorial, 2003; Triggle, 2003; Williams, 2003). Similarly, significant parts of the investment community in the City of London have been very sceptical of the claims of a biotechnology revolution, much to the chagrin of the UK government (Pratley, 2003).

Second, is there an alternative to the biotechnology revolution model? Historical research suggests that major technological changes, such as those produced by the steam engine, the production line or the electric motor, never take place in a vacuum. They typically required complementary technical and organisational innovations that constrained and structured their adoption (David, 1990; Freeman and Louca 2002; Rosenberg, 1979). For example, the diffusion of electricity was hampered by problems with cabling, which was only overcome by innovations in steel production (Freeman and Louca, 2002). As a consequence, it can take a long time, typically 40–60 years, for major technologies to produce benefits that even then can be indirect and difficult to detect (Freeman and Louca, 2002; Rosenberg, 1979). As Hopkins (D.Phil Thesis, University of Sussex, 2004) demonstrates, early twentieth-century developments in genetics took many decades to move from the visions of researchers to clinical fruition. Similarly, Benneworth (2003) has highlighted the neglected role of slow, incremental, low-tech biotech innovation.

Rather than focusing on biotechnology, an alternative model might conceptualise recent changes in terms of a shift from craft-based to more industrialised experimentation. In a range of procedures including genomics, high throughput screening, combinatorial chemistry and toxicology, traditional hand-crafted experiments are increasingly being complemented by automated, miniaturised experiments carried out in parallel on populations of samples with complementary analysis of stored and simulated data (Nightingale, 2000).[2]

Even though this change has made the discovery of small molecule drug targets easier, such improvements cannot be extrapolated to the entire process, because this has only shifted the innovation bottleneck to target validation and clinical evaluation where a much more complex, time-consuming and costly process of research is needed. Such systemic changes have generated a range of novel problems associated with information over-load and statistical quality control, and had led to more inter-disciplinary work, an emphasis on speeding up processes and changing organisational structures to maintain output levels. There are parallels here with the problems that Henry Ford encountered when he industrialised production, and it is therefore sobering to realise that the organisational, technical, managerial and social problems that industrialising production generated took many decades to solve.

Whether or not this industrialisation model is realistic, it is becoming increasingly clear that advances in basic scientific knowledge do not simply lead to new medical technologies. Clinical research occurs in highly complex and poorly characterised systems (the bodies of human subjects) and medical practice draws on multiple sources of knowledge, only some of which are at present reducible to science. As a consequence, biological knowledge derived in the laboratory is not easily translated into useful clinical practices.

This problem is now starting to be recognised by policy-makers. The recent FDA White Paper *Innovation or Stagnation* has explicitly stated that 'Today's revolution in biomedical science has raised new hope for the prevention, treatment and cure of serious illnesses. However, there is growing concern that many of the new basic science discoveries…may not quickly yield more effective, more affordable, and safe medical products for patients' (FDA, 2004). In response, the FDA is advocating much greater emphasis on translational and critical path research focused on the clinical assessment of novel products. Initiatives of this kind that address the real problems facing innovators are to be warmly welcomed.

Conclusions

The data we have presented suggests that it is time to rethink the biotech revolution. Policy-makers need to follow the FDA and move away from an increasingly discredited linear model of innovation that sees new drug and diagnostic products as little more than the application of basic research. Instead, policy needs to address the uncertain, systemic nature of technical change and the very long timescales between advances in basic knowledge and productivity improvements (David, 1990; Freeman and Louca, 2002; Rosenberg, 1979).

The FDA's emphasis on the importance of getting our facts right is a welcome development because unrealistic expectations have had a major impact on government policy. Undoubtedly, some of the policy suggestions are intrinsically good ideas, such as promoting better knowledge transfer between industry, universities and the health care system, but successful policy needs to be based on sound evidence and a sense of proportion. This has not always been the case with biotechnology and there is now a substantial mismatch between the real world and the unrealistic expectations of policy-makers, consultants and social scientists.

While we have hinted at an alternative model we can say very little at present about the long-term prospects for biotechnology and our data is compatible with a range of eventualities. A pessimistic perspective might highlight that the biotechnology revolution has been closely associated with a very reductionist, genetic model of disease (Charkravarti and Little, 2003; Lewontin, 1993) that is increasingly being challenged by explanations that emphasise the interaction between environmental, lifestyle and biological factors across the life course (Drews, 2003). Epidemiologists have already noted how the social distribution of a range of common disorders, such as obesity, stomach ulcers and heart disease, has radically changed in the last century, suggesting that the major determinants of these diseases are social rather than *purely* genetic in origin (Wilkinson, 1997). These environmental factors such as poverty and smoking require comprehensive public health programmes rather than unproven high-tech solutions that are unlikely to be delivered in the short term (Triggle, 2003). This uncertainty about the timing and benefits of biotechnology suggests the need for regular checks against the evidence in order to avoid constructing shared expectations that have little empirical foundation.

Our concern is not the future but the present, and more particularly how current expectations and talk of revolutions help generate the social cooperation needed to deal with the very long-term lead times required to create new medicines. Unrealistic expectations are dangerous as they lead to poor investment decisions, misplaced hope and distorted priorities, and may distract us from acting on the knowledge we already have about the prevention of illness and disease.

Notes

1. Reprinted from *TRENDS in Biotechnology*, Vol. No. 22, Issue No. 11, Paul Nightingale and Paul Martin, 'The Myth of the Biotech Revolution', Pages No. 564–569, Copyright (2004), with permission from Elsevier.
2. That very similar processes are found across a range of technologies, many of which have little connection to biotechnology, suggests further problems with the biotechnology driven causality of the revolutionary model.

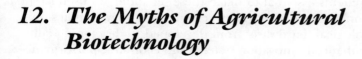

12. The Myths of Agricultural Biotechnology

PETER SENKER AND JOANNA CHATAWAY

Introduction

Agricultural biotechnology is an intensely controversial subject: both protagonists and their opponents accuse each other of promoting myths and denying facts. The impact that these technologies might have on poor farmers in developing countries has been a particularly fertile ground for diverse accounts of what the technology might bring. On the one hand there is the construction of hope-filled mythical biotechnology-based edifices within which farmers will find salvation, and on the other stories about terrifying out-of-control technologies which will rob farmers of seeds, land and ownership. Distinguishing myth and reality is sometimes impossible.

The critical role played by particular organisations, institutions and 'innovation systems' in creating science and technology based products – useful or otherwise for poor farmers and consumers – can get lost in this war of words. Science and technologies do have innate properties, but the institutional context in which they are developed and applied is key to understanding the rate and direction of their development and who benefits from them. This chapter considers agri-biotechnology in the context of observations about the institutions and systems in which it occurs.

Biotechnology is part of a larger set of scientific and technological changes occurring in the 'life sciences'.[1] It can be argued that contemporary life sciences have the potential to contribute in positive ways to efforts to overcome some major agriculture-related problems in developing countries (Conway, 1997; Daar et al., 2002). Conway highlights the potential of biotechnology for addressing conservation as well as productivity in agricultural biotechnology. Daar et al. report on a survey used to identify 10 biotechnologies that could benefit the health of people living in developing countries: examples are genetic modification to produce nutritionally improved crops.

However useful these insights into technological possibilities, there is plenty of scope for myth building. Investing in science is only one key element in developments essential to make these technologies useful. The idea that investing only in agricultural science and technology, or that simply transferring technological 'hardware' from developed to developing countries can result in enhanced innovation systems and economic performance has been attacked from several directions (Byerlee and Fischer, 2002; Clark, Yoganand and Hall, 2002; Douthwaite, 2002; Hall, Yoganand, Crouch and Clark et al., 2004). These authors argue that just because the science exists it cannot be assumed that useful technology will emerge. It cannot be assumed that 'transfer' processes will necessarily occur – from science to useful technology, from North to South, from laboratory to farmer.

Recently a flurry of reports and papers have analysed the shortcomings of the 'linear model' of innovation (e.g., Chataway, Tait and Wield, 2005; Millennium Project, 2004; Oyelaran-Oyeyinka, 2005). In essence, the linear model suggests that investment in discrete areas of scientific endeavour leads seamlessly to technology which can then be transferred to private companies which can then turn the technology into products. But scientific and technical knowledge and product development are outcomes of disciplinary and inter-disciplinary knowledge, and of non-linear interaction between producers and consumers. For useful innovation to occur, knowledge often needs to flow back and forth between producers and consumers. Such reports and the analysis underlying their recommendations suggest that several boundaries need to be crossed: public and private need to communicate and relate, scientific disciplines need to be brought together, and to be related to the management and social studies. This brings into focus the need for highly complex sets of linkages (Gibbons et al., 1994).

The following section looks first at the evolution of biotechnology in Europe and second at some myths about the private sector, biotechnology and the poor. This provides background necessary for understanding the problems which have arisen in the course of efforts towards implementing institutional change designed to create useful biotechnology-related science, technology and innovation.

Biotechnology's fall from grace in Europe

Early applications of biotechnology were designed to appeal to farmers rather than consumers. The lack of demonstrable advantages for consumers contributed to the difficulties of marketing the technology in Europe. The mythology about the revolutionary potential of the technology for feeding the world did not match the reality of products which seemed to extend rather than transform the chemical inputs era of agriculture.

In the late 1980s, the agrochemical industry focused on the production of chemical pesticides to control insect pests, diseases and weeds. There was increased competition in the industry and profits were falling as many major products moved out of their period of patent protection, and there were extensive mergers and takeovers. Most products had been developed with the aim of securing markets in intensive agricultural production of major world crops. The industry developed pesticides which were effective in dealing with crops and pests which were easy to deal with and saturated markets in the developed world. Regulatory pressures increased development costs for new pesticides, and, accordingly, relatively few new ones were brought to market each year. The large multinational companies (MNCs) involved concentrated their efforts almost entirely on developing products for major world commodity crops such as maize, cotton, rice, bananas and soya bean which provided markets large enough to offer them prospects of recouping huge R&D costs. In response to these pressures, managers were looking for ways of escaping from becoming mere producers of commodity chemicals (Chataway, Tait and Wield, 2004). So as to develop highly profitable products, MNCs moved from agro-chemicals to biotechnology. They invested large sums of money in R&D over a period of about 15 years 'to develop products well in advance of any market demand or even public awareness of the technology and its potential benefits and risks' (Tait and Chataway, 2003: 2).

The early products introduced by Monsanto and other companies were based on relatively simple technologies which the companies and publicly funded research laboratories had been working on for some time, so that they could be brought to market quickly. The companies wanted to generate income which could be used to fund R&D on more complex products. They decided to concentrate mainly on herbicide and insect resistance which would help farmers to facilitate crop growth but did not offer any direct benefits to consumers. This was a major factor in European consumers' resistance to the technology. Research on European public attitudes to biotechnology in the late 1980s indicated that attitudes to genetic modification (GM) in Europe as a whole were similar to those reported in Britain. The public wanted to see some benefits to society, not just increased profits for MNCs. A report on a special public debate on GM and the commercial growing of GM crops in Britain found that there was 'strong and wide degree of suspicion about the motives, intentions and behaviour of those taking decisions about GM – especially government and multi-national companies...the debate also highlighted unease over the perceived power of the multi-national companies and of such companies in general' (Department of Trade and Industry, 2003: 6–8).

The public believed that MNCs were motivated overwhelmingly by profit rather than by meeting society's needs and, recognising that their 1990s campaigns failed, companies now acknowledge this. The findings suggest that even when people accepted that there could be benefits of GM technology, they were doubtful whether GM companies would deliver them. Moreover, provision of additional information did not seem to make people feel more favourably towards GMOs. When asked to review their responses to the questions, people readily acknowledged that they did not know much about GM. Although they had anxieties about risks from GM, particularly in relation to the environment and human health, they accepted that GM may offer some benefits. When people in the general population become more engaged in GM issues and chose to discover more about them, they harden their attitudes to GM: the more they discover about GM the more convinced they are that no one knows enough about the long-term effects of GM on human health. In summary, there was little support for the early commer-cialisation of GM crops.

In the context of the developing world, opposition to GM was based less on negative feelings towards GM than on the view that there were better ways to promote development – for example, fairer trade, better distribu-tion of food, income and power, and better government: the public were particularly sceptical about the will of multinational companies to deliver such benefits (Department of Trade and Industry, 2003: 6–8). Campaigns highlighting the potential benefit of biotechnology did little to convince sceptical European consumers.

It is important to note that people's beliefs about biotechnology linked in complex ways to their views about the institutions that are developing them. Evidence or knowledge about the technology is clearly not always the primary factor which influences attitudes and beliefs. For example, public opposition was reinforced by the initial refusal to label GM foods as such when they first appeared on the European market (Tait and Chataway, 2003: 4 and 11).

Myths about the private sector and the poor in developing countries

As already indicated, proponents of agricultural biotechnology promoted the myth that new agricultural technologies – in particular GM foods – were being developed to feed the hungry. During the 1990s some major multi-national corporations claimed that genetic engineering would increase the productivity achieved by farmers in developing countries and alleviate poverty and hunger. One of Monsanto's adverts devised as part of an ill-fated campaign in the late 1990s included the following:

> ...many of our needs have an ally in biotechnology and the promising advances it offers for our future. Healthier, more abundant food. Less expensive crops... With these advances we prosper; without them we cannot thrive. As we stand on the edge of a new millennium, we dream of a tomorrow without hunger....Worrying about starving future generations won't feed them. Food biotechnology will. (Quoted in 'Feeding or Fooling the World-Can GM Crops Really Feed the Hungry?' Report of the Genetic Engineering Alliance 2002: 13, www.fiveyearfreeze.org)

While the rhetoric is appealing and the technology does have potential to increase productivity, Monsanto's investment patterns did not match the sound bites. Research was concentrated in areas where it was thought likely to open up big markets in developed countries - for example, to produce slow-ripening tomatoes – rather than in those which would benefit developing countries such as drought-resistant crops for marginal lands, or foods which have a high nutritional value. In Britain, the Royal Society carried out a study on the use of GM plants for food use and or human health and found no examples of the use of GM technology to improve the nutritional quality of crops and human health (2002: 6). This study confirmed that few of the foods produced so far or being researched and developed were foods which the hungry can afford.

Moreover, small farmers were often badly affected by the advent of GM soya grown and sold by Monsanto – the company's great success story. Total production of soya programmed to be resistant to their Roundup agricultural herbicides, increased by 75% over five years to 2002 and yields increased by 173 per cent. Its use grew rapidly in Argentina partly because it was planted by direct drilling into the soil, helping to solve a problem of soil erosion which had been caused by ploughing on fragile pampas. In the late 1990s, soya became the cash crop for half of Argentina's arable land and a huge cash export providing cattle feed for Europe and elsewhere. But 150,000 small farmers were driven off the land so that more soya could be grown and production of many staples such as milk, rice, maize, potatoes and lentils fell. Some researchers also suggest that GM soya in Argentina is damaging soil bacteria and allowing herbicide-resistant weeds to grow out of control. Soya 'volunteer' plants, from seed split during harvesting, appear in the wrong place and at the wrong time and needed to be controlled with powerful herbicides since they were already resistant to glyphosate. The control of rogue soya resulted in the drift of herbicide spray onto neighbouring small farms and caused them heavy losses in their own crops and livestock and fuelling fears of 'super weeds'. However, as suggested by Monsanto, some of these problems were caused by the growth of the soya as a monoculture and the failure to use the soya as part of an appropriate crop rotation (Brown, 2004).

Biotechnology companies concentrated on a restricted range of crops which offered large and secure markets and involved capital intensive production systems. The transgenic crops which they developed were patented. WTO rules prevent farmers from reproducing patented seeds which they harvest themselves (Altieri, 2005). Efforts by a US company to patent basmati rice caused outcry and highlighted the potential dangers and absurdities involved in new patenting arrangements (Commission on Intellectual Property Rights, 2002). Accordingly, many argue that the application of intellectual property protection to crops may well have negative consequences for poorer farmers (2002).

These developments in biotechnology can be seen as reflecting a longer standing pattern of development of agricultural inputs which have benefited middle income farmers but not poorer farmers. Since the mid-twentieth century, a group of powerful agricultural technologies, including 'Green Revolution' technologies, have been developed by scientists in international research centres, adapted in national research institutions adopted by extension agencies and agro-chemical and seed companies and marketed to farmers. These technologies include uniform high-yield crops, mechanical and energy inputs and synthetic chemicals. They tend to reduce indigenous biodiversity and are not designed for small resource poor subsistence farmers. They have benefited middle income farmers in some developing countries and many claim that the overall development impact in specific contexts has been positive. Without institutional reform GMO technology is likely to continue as part of a complex system of international legislation and trade restrictions which may tighten corporate control over food production without benefits to small farmers (Senker, 2000).

For example, in India, vast state expenditures on the Green Revolution were concentrated in the Punjab – the most fertile area of the country. Food grain production there increased from about 3 million tons in 1965–1966 to over 25 million tons in 1999–2000. The Punjab – a state with 24 million people – about 2 per cent of the population – produces over 12 per cent of India's food. But three quarters of Indian farmers living in poorer states without access to large area of land were marginalised and failed to benefit from the Green Revolution. Even in the Punjab, the number of smallholdings dropped sharply because many farmers could not afford the necessary irrigation or fertiliser (Patel, 2007: 125–126). In contrast, since 1957 the state of Kerala has adopted political solutions to agricultural and social problems, rather than adopting technological fixes such as the Green Revolution or biotechnology. The Kerala government's policies of land redistribution, food distribution, employment guarantees, health care and education resulted in the highest literacy, health and general welfare in

India. Indeed, despite the poverty of Kerala's 30 million population, literacy levels and life expectancy are higher on average than in some parts of the United States. Moreover, the Keralan approach has provided more enduring benefits than some Green Revolution approaches. In the 1990s, 20 years after the Green Revolution, while malnutrition seems to have increased almost everywhere else in India, indicators of health and welfare remained high in Kerala (126 and 127).

Nevertheless, new initiatives such as 'The BiOS Initiative – the Biological Innovation for Open Society' (www.bios.net) often referred to as 'open source biotechnology' may be important in overcoming institutional constraints to the benefit of smaller farmers. Open source software, such as Linux, has had a dramatic impact in the IT sector and it is hoped by its promoters that BIOS may contribute to breaking up the concentration of powerful commercial interests in agricultural biotechnology. BIOS is hosted by CAMBIA, a non-profit organisation in Australia and the International Rice Research Institute (IRRI), one of the Consultative Group for International Agricultural Research (CGIAR) centres.

Anti-GM mythology – 'Frankenfoods'

Several authors responded to anti-GM campaigns and criticisms of the way in which MNCs have developed GM technology by pointing out that GM critics themselves develop mythologies. For example, the International Food Information Council (IFIC) lists and purports to evaluate 22 myths disseminated by opponents of biotechnology and charges opponents of the technology of basing arguments in value laden ideological commitments rather than examination of evidence (IFIC, 2003). Articles with headlines referring to 'Frankenfoods', clearly grounded in scare tactics rather than solid evidence, made this charge an easy one to make; the mythologising about the benefits of the GM were clearly being matched and trounced by a backlash based very clearly on mythological creatures, conspiracy theories and plays on fears of 'out of control' substances devouring human life. Such campaigns were orchestrated largely by environmental NGOs, deeply opposed to GM technology and using sophisticated communications strategies to get their messages across.[2]

In a book which looks critically at the way that environmental groups develop their arguments regarding science and technology, Dick Taverne (2005) portrays environmental NGO opposition as a matter of faith rather than reason:

> Genetic modification is to Greenpeace, Friends of the Earth, and kindred organisations, what abortion is to Roman Catholics and American evangelicals... To many of the Green lobbies rejection of GM technology has become a tenet of faith, and any evidence that contradicts the faith is simply irrelevant.

Taverne quotes Lord Melchett then Director of Greenpeace responding to a question about the nature of his opposition to GMOs:

> Question: Your opposition to the release of GMOs, that is an absolute and define opposition...not one that is dependent on further scientific research?
>
> Lord Melchett: It is permanent and definite and complete opposition. (132)

With regard to developing countries, Taverne considers arguments made by some environmental and development lobbies to be reflections of a set of values and deep-rooted ideologies: these are not shared by practitioners and policymakers in the global South as is often suggested by NGOs (127).

Broad questions about the nature of scientific truth clearly lie outside the scope of this book and continue to be discussed in depth (e.g., see Beck, 1992; Jasanoff, 2005; Wynne, 1996). The rationales of opposition and support for biotechnology have also been widely discussed (e.g., see Rayner, 2003). As we have already indicated, views about whether or not biotechnology has anything to offer developing countries diverge widely, and there is a complex mixture of interest based and value based opposition.[3] For example, the Neem Campaign tries to protect indigenous knowledge systems, working on behalf of small holder farmers and others (interest-based opposition) and also work to protect resources of the Third World from MNCs who wish to exploit these resources for the benefit of their largest markets in the developed world (Shiva, 1997). This opposition is based on protecting interests, but group activities are rooted in deep belief systems.

Sir Robert May, then British government's chief scientist, was reported as saying:

> It seems unclear, at this point, whether GM crops have the potential to be a further notch up in (agricultural) intensification and as such, not good. But equally, they have the potential to enable us to redesign crops so that we work with nature and shape the crops to the environment rather than shape the environment to the crops in ways which are unsustainable.... I take a rather different view of this from Greenpeace, who see it as a prime campaign issue and necessarily bad. (May, quoted in Douglas, 1999 – see Senker, 2000: 214–216)

May's views are based on considered opinion but are also rooted in his position as a leading scientist and protagonist of a set of scientific methods.

We have argued that biotechnology is rooted in a set of institutions. The nature of genetic engineering is not entirely determined by these institutions but the direction and rate of innovation cannot be understood without considering them. In the following section we look more in more detail at the institutions which influence the ways in which technologies are and are not developed for use in developing countries.

Understanding the institutional context

There are severe problems involved in making science and technology work for the poor in developing countries (Byerlee and Fischer, 2002; Clark, Yoganand and Hall, 2003). Long-standing issues are compounded by trends associated with the evolution of biotechnology and genomics. They include a relative increase in the share of R&D being undertaken by the private sector with associated intellectual property protection acting as an additional barrier to small farmer access; a decrease in public sector funding for agricultural R&D (Krattiger, 2002) and the emergence of new international and regional risk and trade regulations (Keeley and Scoones, 2003). Investment trends in R&D are particularly worrying perhaps as poorer developing countries are highly dependent on public sector research (Cohen, 2005).

Concurrent with these changes in context and in some respects intimately connected to them, a renewed and powerful critique of traditional models of agricultural research is emerging. The entire budget of the CGIAR the principal international institution that fosters developing country agricultural research around the world, is less than the amount each of the top four agro-chemical and biotech companies spend on their own R&D every year (Spielman and Grebmer, 2004).

How could biotechnology help the poor?

Hunger is generally not a consequence of failures of food production. For example, starvation, mass hunger and hunger-related deaths in Africa have resulted from numerous factors such as armed conflict and drought rather than because appropriate crops have not been available to be planted (Patel, 2007: 148–163). Nevertheless, if the right circumstances can be created, there may be significant possibilities for biotechnology to help the poor, but there are critical decisions to be made about the resources and capacity building efforts needed to exploit them (Chataway, 2005). To what extent should funds be allocated to building scientific capabilities and to what extent should resources be devoted to the range of institutions and organisations which sustain technological capacity, such as regulatory and policy making bodies, communications and networking groups, market support mechanisms?

Hall and others suggest that if capacity and enhanced production capabilities are to result from investment in agricultural R&D attention needs to be given to the system of production and innovation as a whole; the policy environment and processes involved in learning (2004). Analysis based on a linear model in which scientists offer farmers solutions is inadequate as the foundation of sustainable and beneficial innovation systems. A model in which farmers, NGO groups and a range of other relevant actors, including appropriate private and public sector players, are involved as part of a process of scientific and technological innovation is essential and institutional innovation must allow for this type of communication to occur and to be meaningful.

Change in agricultural R&D is driven partly by macro and international economic and political shifts determining, for example, the level and nature of private and public sector investment globally in agricultural R&D, the type of international patent and risk regulation producers have to abide by. It is also driven partly and more purposively as part of national policy processes. The interaction between international and local policy dynamics influence local technology development in multiple ways.

'Emerging economies' are investing heavily in the technology and, although some analysts worry about detrimental environmental impacts, there is now evidence that farmers are benefiting from GM crops and other agriculture related biotechnology developments in countries such as China, Argentina and South Africa (Cohen and Paarlberg, 2004; Thirtle, 2003). These countries are making substantial investments in the development of biotechnology and the integration of new life science based products and techniques into national farming systems, although even in some cases such as Argentina considered earlier, outcomes even in such economies are not universally entirely favourable.

The situation in poorer countries and amongst smaller and poorer farmers seems less clear. For countries that have very limited resources to invest in the technology and to develop their regulatory and policy systems, the options for developing and adapting technologies are restricted (Cohen, 2005). Whilst, for example, GM technologies that can offer increased drought resistance in crops or non-GM processes that can speed up development of improved varieties potentially have much to offer poorer farmers, a number of constraints inhibit the useful application of such technologies. Capacities to adapt the technology to local problems and to create policies that correspond sufficiently well to new complexities associated with the technologies are often lacking. Technology transfer and capacity building efforts often seem inadequate. Although progress may be made in laboratories, farmers, particularly small farmers, do not reap rewards (Clark, Yoganand and Hall, 2002, Horstkotte-Wesseler and Byerlee, 2000). However, Hall discusses a programme called the Andhra Pradesh–Netherlands Biotechnology Programme (ABNBP) which does appear to have yielded some positive results. Innovation within this programme requires institutional flexibility and iterative and ongoing involvement of farmers and scientists in addressing problems. Many research partnerships fail because they are embedded in structures that separate 'technical' from 'social and economic' problems and separate researchers from producers and users. Problems can often only be understood and resolved from a more holistic perspective that acknowledges the context, specific nature of problems and solutions and the institutions which could resolve them. The Andhra Pradesh–Netherlands is a good example of a system based on a 'bottom-up' approach.

Public-Private Partnerships

Recent reports are concerned to protect agricultural research aimed at poor farmers and to make sure that research can be restructured to serve those farmers better. They focus on the construction of systems of innovation which could serve the poor rather than simply on investment in scientific and technical skills. They also highlight the role that public-private partnerships (PPPs) might play. Efficient biotechnology and genomics related agricultural research can require sophisticated scientific and technical equipment highly qualified scientists. Public-private partnerships (PPPs) may be increasingly important in using technology to help the poor. The case for large international PPPs rests broadly on the following logic.

Most of the new bio-technologies developed by the private sector are designed for industrialised country markets. Thus partnerships between the research-intensive companies that hold patents and public sector or multilateral agencies concerned with developing country priorities can form useful tools for converting technologies to meet the needs of poorer farmers and developing country markets (Byerlee and Fischer, 2002). Large companies may engage on the basis of a market segmentation principle: in countries where there is no commercial interest for the innovating company, technology is donated for use in domestic crops. A variety of agreements of this sort have been reached. For example, Monsanto has entered into an agreement with the Kenyan Agricultural Research Institute (KARI) so that a Monsanto transgene for control of African sweet potato virus disease can be used without charge in all of Africa (2002). Another example is Syngeneta's donation of delayed ripening technology for use in several Asian countries. In such cases developing countries may be able to use the latest technologies to solve local problems.

Several other arguments support the establishment of PPPs including the observation that not only is the public sector declining relative to the private in agriculture research investment, but the public sector also now behaves more and more like the private sector. In both private and public sectors, the rate of patenting has increased enormously (Dutfield, 2004). Developing countries should not rule out engagement with international private sector companies on the assumption that they will necessarily get fairer treatment from the public sector. Moreover, the public sector has well observed weaknesses in delivering goods and, as some authors have pointed out, the private sector may be a useful force in creating a more user friendly research service (Gibbons et al., 1994).

But the success of biotech related PPPs has been patchy. There are far fewer PPPs than might initially have been expected and the participation of the large multinationals which have been investing so heavily in the technology is not yet very significant (Spielman and Grebmer, 2004). Also, as noted above there are a small number of private sector actors. There is only one major multi-national company involved in that country's PPPs, Monsanto (Ayele and Wield, 2005). Monsanto's involvement is based on technology donation and the company has also trained Kenyan scientists and supported some of the NGO groups which have sprung up around biotechnology products. Three Kenyan and South African based private sector companies are active in Kenya. Research in Kenya suggests that PPPs designed to support the development of TC bananas have made substantial contributions to technological capacity building. However, initially small farmers had difficulty in coping with TC bananas (Smith, 2004). Their difficulties related to the fact that the TC bananas all arrived at one time and local markets became flooded. Quotes from interviews with small farmers carried out for this research included the following: 'TC bananas are not meant for local cultivation', 'Kenya needs some mechanism to add value to its bananas', 'No one thought ahead about surplus bananas'. Lack of private sector engagement in developing regional or export markets was therefore seemingly a problem in the Kenyan context. The recent development of local private sector capacity may, however, resolve current difficulties (Karembu, 2004).

Ayele and Wield (2005) examined the contributions made by PPPs in two other very different contexts. In line with observations about the TC bananas projects, one of their general conclusions is that PPPs contribute to science and technology capacity building but do not always involve producers and users at an early stage. This limits their effectiveness for poorer farmers. However, Ayele and Wield also suggest that the will to develop PPPs and to encourage involvement of the private sector matters in terms of the types of systems that get created; PPPs become significant as 'systems builders' and this may be important in terms of longer-term development.

PPPs in health and agriculture: Understanding the differences

In health, PPPs are having a major impact on the development for vaccines for neglected diseases. In agriculture, PPPs' scale of operations seems much more restricted. Researchers working on projects associated with the ESRC Innogen research centre have analysed both. And their results have indicated clear differences between agricultural and health biotechnology. One of the main differences is that market structures in health and agriculture are very different with knock-on effects for innovation in both areas. In health the target areas for PPPs, major neglected disease areas, for example, potentially constitute large undifferentiated markets whereas the targets in agriculture in poorer developing countries are embedded in small activity in local and regional markets. In health large global PPPs (GPPPs) have emerged in major disease areas but there are very few equivalents in agriculture.

For example, Golden Rice – vitamin A enriched rice – might be thought of as a project that is more akin to health GPPP programmes: a very large international programme with a potential global reach. However, trying to adapt the process of vitamin A enrichment to local varieties is enormously complex and adds very considerably to the costs involved in the project (Drake, 2005). The recently initiated 'Harvest Plus' programme of biofortification projects, based at CGIAR centres, may be an attempt to model more large agri-global PPPs in the mould of health programmes. But it is hard to see how these projects will avoid the need for very significant local adaptation and a very different set of interactions with local farmers and consumers. A few global PPPs in agriculture may emerge, but the majority will have a more restricted geographical focus. This will mean that their resource requirements and structure will be quite different.

Local and traditional knowledge are key to agricultural innovation in developing countries and will therefore be important to the PPPs set up to work in this sector. Intense interaction with local communities and local civil society is also important to health PPPs but not in the same way. In health the interaction is about explaining what a vaccine could do and engaging on the basis of good ethical practice, raising awareness and demand, making sure that products are needed and wanted: indeed, these are requirements for successful innovation in both areas (Chataway, Smith and Weild, 2005). However, in agriculture innovation needs to take place on the basis of local knowledge, cultural preference and local environmental conditions. Such considerations indicate that PPPs need to be structured according to the characteristics of the sectors they are working in.

Conclusions

We do not suggest that all claims to truth are equally valid, but it is important to recognise that policy-based arguments are often rooted in a complex mix of interest and value-based perspectives. Indeed, Tait (2004) and others suggest that we accept the natural limitations of definitive truths but use and scrutinise 'evidence-based' assessment in trying to find our way through the maze of truth claims. This is what we have tried to do in this chapter.

It is a myth that agricultural development, and in particular genetic modification, is mainly driven by the motivation of feeding the hungry. Moreover, hunger is more often the consequence of factors such as war, drought and land distribution rather than of lack of availability of suitable crops. Large MNCs have driven the principal technological changes which have affected agriculture in the last several decades. Their aim has been to make profits from agriculture both in the developed world and in developing countries. A common factor in the policies they have pursued to exploit the products they have developed is to secure the largest possible markets for their products , and a common factor in the routes they have followed to secure this goal has involved the worldwide diffusion of monocultures. The history of agriculture shows that monoculture tends to make plant diseases, insect pests and weeds more severe, and that intensively managed and genetically manipulated crops soon lose genetic diversity. This indicates that resistance to transgenic crops will evolve among insects, weeds and pathogens as has happened with pesticides. Studies of pesticide resistance demonstrate that unintended selection can result in pest problems that are greater than those that existed before deployment of novel insecticides. Diseases and pests have always been amplified by changes toward homogeneous agriculture.

Nevertheless, much of the opposition to agri-biotechnology by NGOs is as reliant on myth as are the claims made by MNCs. And there seems to be some potential for agri-biotechnology to contribute to the improvement of food availability to the hungry – for example, by means of PPPs.

Notes

1. Very broadly the term 'life sciences' includes any study or discipline that contributes to our understanding of life organisms and processes and relationships to one another and to their environments.
2. See also Chapter 4 in this volume.
3. Categories of 'value based' and 'interest based' positions have been developed by Bruce and Tait (2003).

Epilogue

Myths have been an important and consistent element of human communities. Tribes of hunter-gatherers, the city states of ancient Greece and Rome, and feudal societies all had well-developed mythologies which contributed to the cohesion of their communities. Mythologies provided explanations and understandings of everything from social structure to the workings of the solar system, as well as providing roles, identities and rituals which governed everyday life and managed social and environmental change.

Despite the huge growth in scientific knowledge over the past 500 years we have demonstrated that, even in the twenty-first century, society still relies heavily on myths. Capitalist society relies on mythology for its very existence. Myths about markets and their operation, as well as the actors and agencies which bring them to life continue to abound, appearing with remarkable consistency with each wave of innovation.

The struggle for theory to displace myth continues to be an important intellectual enterprise, found in all the faculties of a typical university, including Business Schools and faculties of computing and engineering, as well as those of Social Sciences, Humanities, Media and the Arts. Each wave of human innovation provides more data to be worked over by the machinery of academic, scientific and artistic enterprises, while operators in the market (which includes some of the above) look for opportunities to commercialise innovation in thought and deed by transforming innovation into products and services which can be sold into open and niche markets.

The waves of new technology have provided both a rich source of inquiry and have lent themselves to immediate and in many cases highly effective commercialisation. New technology has kept the capitalist enterprise in business, providing a rich bank of novelties which are claimed to solve all manner of communication, information and technological problems and bringing new myths and theories which explain their place in our world and generate a unique and complex mix of insight, understanding, knowledge and myths. The gap between myth and theory sometimes becomes as blurred as the lines between knowledge producers. Once, R&D

was conducted within the silos of the different sectors of university and military establishments on the one hand, and private corporations on the other. But the lines between these have become blurred and porous. Scientific breakthroughs, with the vast expense which they sometimes incur, are as likely to come out of private foundations as the university establishment, which in our critically engaged view we think is sometimes as likely to produce myths as theory! The new technology innovation curve has destabilised not only what we think we know about the world, but some of the processes by which such knowledge is produced. Legacy systems which managed innovation in social life are challenged by the kinds of innovation discussed in this book. Intellectual property rights are a good example, as are innovations in social control such as enhanced surveillance systems based on the new technologies. The control systems of police, security forces, traffic wardens and doormen found on the ground have been supplanted by a wide range of technological systems including the now ubiquitous ones such as CCTV, with its promise of infallibility – the camera never lies. Innovations in social life include new social identities and lifestyles constructed around a range of social categories such as health and illness. New technologies are deployed in both the social construction and the actual lived experience of such categories providing new avenues of inquiry and insight. These new forms of data tell us new things about the world yet create uncertainty in the conclusions which might be reached – is the rate of traffic offences rising or merely captured on film more often? Does the rate of traffic offences fall because drivers adjust their behaviour by driving more carefully, or by driving equally badly but down a road with no cameras? Do drivers really drive with a lower level of skill than what prevailed prior to the introduction of driving tests?

The innovation curve often reveals the extent of that which we did not know before, might turn orthodoxy on its head, and generally upset our most dearly cherished beliefs about the world, including those which previously we have been pleased to regard as relatively stable knowledge. Furthermore, people who wish to exploit a new technology always need to attract the resources necessary to develop and apply it, and they might create and disseminate claims which become myths to help them acquire those resources, sometimes attracting millions of investment on what is later found to be lost causes – endeavours subsequently disproved as nonsense such as those associated with the so-called millennium bug. Meanwhile, those who oppose and fear new technologies also use myths to support their case. However, while scientific inquiry may require a different kind of rigour than that required to raise the funding for it, we do not underestimate the difficulty of generating theory and new knowledge in place of mythology. Indeed, a core problem in intellectual enterprise is our uncertainty over which is which, new knowledge and understanding versus myth. The apparent instability of our knowledge of the world, which comes with each disproof and paradigmatic shift, may lead us to assume that a degree of uncertainty is in scientific terms a necessary precondition of our inquiry. Yet to demonstrate uncertainty in doing deals worth millions of dollars in investment might undermine such partnerships.

The more colourful the myths used to promote new technology, the greater the exaltation of the benefits – including stories about how the new technology will benefit the poor as well as the rich, how it will improve the environment, how it will put an end to human suffering and so on, the more power its promoters acquire to attract the funds necessary to keep the show on the road. Capitalist enterprise largely supports the prevalent myth that the continuation and expansion of capitalism throughout the world are essential to human well-being, that there is no alternative, or even modification or version which might allow different kinds of social activities including different kinds of innovation curve to occur.

As each new major technology cluster is envisaged, developed and applied, similar patterns seem to emerge. The initial protagonists of each new technology forecast that it will have considerable benefits to all humankind as well as market, communities and environment. In order to obtain funding, reputation and esteem, they must envisage that applications will be developed and applied much more quickly than happens in the event, and that their benefits will be more evenly diffused between rich and poor than happens in practice. Development and its diffusion is not under the developers' control; such processes occur through the filtering of social mechanisms, structures and discourses which irrespective of the intentions or activities of developers act in powerful and sometimes unpredictable ways. Rather as the author of a new social theory cannot control what happens to it once it is released into the complex systems of the world, innovators take their chances in capitalism in its entirety. They face limits in their ability to pick and choose the aspects with which they will deal, and even the most powerful of the world's monopolies in time will face new upstarts, pretenders to their throne, who following the logic of innovation curves will manifest, with their own motivations, schemes and alliances.

This book includes numerous interesting and useful findings from an extensive range of diverse studies, drawing on an equally wide range of disciplinary perspectives. We suggest that this should help readers to avoid falling into the trap of adopting narrow approaches to the issues they want to study or in overlooking the complexity of what we have argued to be a *social* process as much as a technological one. Indeed, we believe that there is considerable evidence that the adoption of unduly narrow approaches in the name of rigour reduces the value of at least some of the work published today.

Each chapter in this book considers aspects of its broad themes – myths, technologies, innovation and inequality. They are written from various different disciplinary and multi-disciplinary perspectives. They demonstrate the huge variety of the myths of technology and how far and wide their tentacles stretch: they reach into our everyday lives on the one hand, and yet shape our world views and influence official institutions and powerful organisations on the other. The struggle for theory and knowledge over ideologies and myths is as alive as it ever was, and we invite readers to embrace it as a positive and indeed necessary part of our most healthy endeavours to make sense of the world in which we live.

Judith Burnett, Peter Senker and Kathy Walker

Glossary

AI	Artificial Intelligence
AIDS	Acquired Immunodeficiency Syndrome (also see HIV and HIV/AIDS)
BiOS	Biological initiative for Open Society
CAMBIA	originally stood for Center for the Application of Molecular Biology to International Agriculture
CCTV	Closed Circuit Television
CD	Compact disc
CDA	Crime and Disorder Act, 1998 (UK)
CGIAR	Consultative Group on International Agricultural Research
DARPA	Defense Advanced Research Project Agency (USA)
DDI	Didanosine
DiTV	Digital interactive Television
DoH	Department of Health (UK)
DOT force	Digital Opportunities Task force
DOT-COM	Digital Opportunities through Technology & Communication Partnerships (USA)
DOT-GOV	Digital Opportunity through Technology: Government (USA)
DVD	Digital video disc, or Digital Versatile Disc
EPG	Electronic Programme Guide
ERP	Enterprise Resource Planning
ESRC	Economic and Social Research Council (UK)
FDA	Food and Drug Administration (USA)
FRCCTV	Facial recognition closed circuit television
G-7	Group of Seven Finance Ministers from industrialised countries
G-8	Group of Eight: an international forum for the governments of Canada, France, Germany, Italy, Japan, Russia, the United Kingdom and the United States.

GDP	Gross Domestic Product
GII	Global Information Infrastructure
GM	Genetic Modification
GMOs	Genetically Modified Organisms
GPPP	Global Public-Private Partnerships
HAART	Highly Active Anti-Retroviral Treatment
HEI	Higher Education Institution
HIV	Human Immunodeficiency Virus (also see AIDS)
HIV/AIDS	Human Immunodeficiency Virus / Acquired Immuno-deficiency Syndrome
I/O	Input/output
ICTs	Information and Communications Technologies
IRRI	International Rice Research Institute
ITC	Independent Television Commission (UK, superseded by Ofcom, the Office of Communications in December 2003)
IEEE	Institute of Electrical and Electronics Engineers (USA)
IFIC	International Food Information Council
IGBiS	Institute for the Study of Genetics, Biorisk and Society, University of Nottingham, UK
IP	Intellectual Property
ISAD	Information Society and Development
IT	Information Technology
LEO	Lyons Electronic Office (UK)
ITU	International Telecommunications Union
Mabs	monoclonal antibodies
MNC	Multinational Corporation
MUD	multi-user dungeon
NII	National Information Infrastructure (USA)
NGO	Nongovernmental organisation
NHS	National Health Service (UK)
NTIA	National Telecommunications and Information Administration (USA)
OECD	Organisation for Economic Co-operation and Development
OECD ANBERD	Analytical Business Enterprise Research and Development
OSI	Open Systems Interconnection
PC	Personal computer
PhRMA	Pharmaceutical Research and Manufacturers of America
PPP	Public-Private Partnerships

PTZ	pan, tilt and zoom
R&D	Research and Development
RAM	Random accesss memory
SAGE	Semi-Automatic Ground Environment command and control air defence system (USA)
SDLC	Systems development life cycle
SEO	Search Engine Optimisation
SMS	Short Message Service (SMS), often called text messaging
SPRU	Science and Technology Policy Research Unit, University of Sussex, UK
SSEC	Selective Sequence Electronic Calculator (USA)
TCP/IP	Transmission Control Protocol and Internet Protocol. (A suite of data communications protocols.)
UN	United Nations
UNAIDS	Joint United Nations Programme on HIV/AIDS
UNESCO	United Nations Educational, Scientific and Cultural Organisation
UEL	University of East London, UK
USAID	US Agency for International Development
USPTO	US Patent Office
VCR	Video Cassette Recorder
VP	Vice President
WPINDEX Derwent	World Patents Index
WSIS	World Summit on the Information Society, Geneva, Switzerland
WIPO	World Intellectual Property Organisation
WTO	World Trade Organisation

Bibliography

Adorno, T. (1975), Culture Industry Reconsidered, in P. Marris and S. Thornham (Eds, 1996) *Media Studies: A Reader*. Edinburgh: Edinburgh University Press.

Aeschylus (1990), *Prometheus Bound*. Oxford: Oxford University Press.

AIDSmap (2003), Undetectable Viral Load, AIDSmap, www.aidsmap.com.

Alaimo, S. (1994), Cyborg and Ecofeminist Interventions: Challenges for an Environmental Feminism, *Feminist Studies*, 20 (1): 133–152.

Alexa: www.alexa.com a Google-owned free site providing daily details of site traffic for the top world web sites.

Allen, D.K. and Wilson, T. (1996), Information Strategies in UK Higher Education Institutions, *International Journal of Information Management*, 16 (4): 239–251.

Allen, D. and Kern, T. (2001), Enterprise Resource Planning Implementation: Stories of Power, Politics, and Resistance, in N. Russo, B. Fitzgerald and J. DeGross (Eds) *Realigning Research and Practice in Information Systems Development*. Boston: Kluwer Academic Publishers.

Altieri, M.A. (2005), *Some Ethical Questions: The Myths of Agricultural Biotechnology*. Paper presented at National Catholic Rural Life Conference, March 2005. Available at http://www.ncrlc.com/altieri_myths.html 15

Anderson, K. (2001), The Nature of Race, in N. Castree and B. Braun (Eds) *Social Nature*. Oxford: Blackwell.

Andrews, A. (2006), Big Brother Duo Look to Cash in on Appetite for Interactive TV, *The Times*, 7 April 2006.

Angell, I. and Ilharco, F. (2004), Solution is the Problem: A Story of Transitions and Opportunities, in C. Avgerou, C. Ciborra and F. Land (Eds) *The Social Study of Information and Communication Technology: Innovation, Actors, and Contexts*. pp. 38–61. Oxford: Oxford University Press.

Armes, R. (1988), *On Video*, London: Routledge

Armitage, R. (May 2002), *To CCTV OR NOT CCTV?*, NACRO, online. Available at http://www.epic.org/privacy/surveillance/spotlight/0505/nacro02.pdf

Arundel, A. and Mintzes, B. (2004), *The Impact of Biotechnology On Health*, Merit-INNOGEN Working Paper, INNOGEN, University of Edinburgh-Open University.

Ayele, S. and Wield, D. (2005), Science and Technology Capacity Building and Partnership in African Agriculture: Perspectives on Mali and Egypt, *Journal of International Development*, 17 (5): 631–646, published online 23 June 2005.

Bainbridge, D. (1999), *IP*. Harlow: Pearson Education Limited.

Balsamo, A. (1996), *Technologies of the Gendered Body: Reading Cyborg Women.* London: Duke University Press.

Bangsberg, D., Charlebois, E., Grant, R., Holodniy, M., Deeks, S., Perry, S., Conroy, K., Clark, R., Guzman, D., Zolopa, A. and Moss, A. (2003), High Levels of Adherence Do not Prevent Accumulation of HIV Drug Resistance Mutations, *AIDS*, 17 (13): 1925–1932.

Barbrook, R. (2007), *Imaginary Futures: From Thinking Machines to the Global Village.* London and Ann Arbor: Pluto Press.

Barthes, R. (1972), *Mythologies.* London: Jonathan Cape.

Bazalgette, P. (2001), Big Brother and Beyond, RTS Huw Wheldon Memorial Lecture 2001. Available at http://www.rts.org.uk/lectures_det.asp?id=2153&sec_id=354. (accessed on 25 October 2004).

Bazalgette, P. (2005), Billion Dollar Game: How 3 Men Risked It All and Changed the Face of TV, *London: Little, Brown Book Group.*

BBC (2005), *From the Cottage to the City: the Evolution of the UK Independent Production Sector.* An Independent Report commissioned by the BBC, September 2005. London: BBC.

BBC News (2005), *Google Restarts Online Books Plan.* BBC News 01/11/05. http://news.bbc.co.uk/1/hi/technology/4395656.stm. 25/05/08

Beck, U. (1992), *Risk Society: Towards a New Modernity.* London: Sage.

Bell, D. (1973), *The Coming of the Post-Industrial Society: A Venture in Social Forecasting.* New York: Basic Books.

Bell, J. (1998), The New Genetics in Clinical Practice, *BMJ*, 316: 618–620 (21 February)

Bell, J.I. (2003), The Double Helix in Clinical Practice, *Nature*, 421: 414–416.

Benholt Thompsen, V. and Mies, M. (1999), *The Subsistence Perspective: Beyond the Globalized Economy.* London: Zed.

Benneworth, P. (2003), Breaking the Mould: New Technology Sectors in an Old Industrial Region, *International Journal of Biotechnology*, 5 (3/4): 249–268.

Benton, T. (1993), *Natural Relations: Ecology, Animal Rights and Social Justice.* London: Routledge.

Berlin A., Brettler, M.Z. and Fishbane M. (2003), *The Jewish Study Bible: Tanakh Translation, Torah, Nevi'Im, Kethuvim.* Oxford: Oxford University Press.

Beynon-Davies, P. (2002), *Information Systems: An Introduction to Informatics in Organisations.* Basingstoke: Palgrave Macmillan.

BHIVA (2005), *Draft BHIVA Treatment Guidance.* London, British HIV Association.

Bioscience, Innovation and Growth Team (BIGT) (2003), *Bioscience 2015.* London: DTI, HMSO.

Birke, L. (1994), *Feminism, Animals and Science.* Buckingham: Open University Press.

Birke, L. (1999), *Feminism and the Biological Body.* Edinburgh: Edinburgh University Press.

Black, S.A. and Porter, L.J. (1996) Identification of the critical factors of TQM. *Decision Sciences*, 27(1) 1–27.

Blaikie, P. (2001), Social Nature and Environmental Policy in the South, in N. Castree and B. Braun (Eds, 2001) *Social Nature.* Oxford: Blackwell.

Boily, M., Godin, G., Hogben, M., Sheer, L. and Bastos, F. (2005), The Impact of the Transmission Dynamics of the HIV/AIDS Epidemic on Sexual Behaviour: A New Hypothesis to Explain Recent Increases in Risk Taking-Behaviour among Men Who Have Sex with Men, *Medical Hypotheses*, 65 (2): 215–226.

Bordo, S. (1993), Feminism, Foucault and the Politics of the Body, in C. Ramazanoglu (Ed.) *Up Against Foucault*. London: Routledge.

Bourdieu, P. (1984), *Distinction: A Social Critique of the Judgment of Taste*, Trans. R. Nice. Cambridge, MA: Harvard University Press.

Bowers, J. and Bradac, J. (1982), Issues in Communication Theory: A Metatheoretical Analysis, in M. Burgoon (Ed.) *Communication Yearbook* Vol 5, pp.1–27, Beverly Hills, CA: Sage.

Braidotti, R. (1989), Organs without bodies, *Differences*, 1: 147–161.

Braidotti, R. (2002), *Metamorphoses: Towards a Materialist Feminist Theory of Becoming*. Cambridge: Polity.

Braidotti, R. and Lykke, N. (1996), *Between Monsters, Goddesses and Cyborgs*. London: Zed.

Brown, N. and Michael, M. (2003), A Sociology of Expectations: Retrospecting Prospects and Prospecting Retrospects, *Technology Analysis & Strategic Management*, 15 (1): 3–18.

Brown, P. (2004), GM Soya 'Miracle' Turns Sour in Argentina, *The Guardian*, 16 April.

Brown, S, (1998), What's the Problem, Girls? CCTV and the Gendering of Public Safety, in C. Norris, J. Moran and G. Armstrong (Eds.) *Surveillance, Closed Circuit Television and Social Control*, chapter 11. Aldershot: Ashgate Publishing Ltd.

Bruce, A. and Tait, J. (2003), *Interests, Values and Biotechnological Risk*, INNOGEN Working Paper No.7, December, London: ESRC.

Bryld, M. and Lykke, N. (1999), *Cosmodolphins: Feminist Cultural Studies of Technology, Animals and the Sacred*. London: Zed.

BT Redcare January (2003), online http://www.btredcarevision.com/pdf/Vision_ ReducingCrime_Jan2006.pdf

Bulkley,K. (2003), 'I'm a celeb, have a bet on me', *Media Guardian*, 19 May 2003. Available at http://www.katebuckley.co.uk/imacelebguardian.html. (accessed 25/10/04).

Butler, D. and Butler, G. (2000), *20th Century British Political Facts 1900–2000*. London: Macmillan.

Byerlee, D. and Fischer K. (2002), Accessing Modern Science: Policy and Institutional Options for Agricultural Biotechnology in Developing Countries. *World Development*, 30 (6): 931–948.

Cain, P.J. and Hopkins, A.G. (2002), *British Imperialism 1688–2000*. London: Longman.

Cairncross, F. (1997), *The Death of Distance: How the Communications Revolution Will Change Our Lives*. London: Orion.

Callon, M. (1986), Éléments pour une sociologie de la traduction. La domestication des coquilles Saint-Jacques et des marins-pêcheurs dans la baie de Saint-Brieuc, *L'année Sociologique*, 36, 169–200.

Caminer, D. (2003), Behind the Curtain at Leo, in *IEEE Annals of the History of Computing*, April–June 2003, 25 (2), Los Alamitos, CA: IEEE Computer Society.

Caminer, D.T., Aris, J., Hermon, P. and Land, F. (1996), *User-Driven Innovation: The World's First Business Computer*. London: McGraw Hill Book Company.

Cassidy, J. (1997), The Next Thinker: The Return of Karl Marx, *The New Yorker*, 20–27 October, 248–259.

Castells, M. (1996), *The Rise of the Network Society. The Information Age: Economy, Society and Culture,* vol I. Oxford: Blackwell Publishers.

Castells, M. (1998), *End of the Millennium. The Information Age: Economy, Society and Culture,* vol III. Oxford: Blackwell Publishers.

Castells, M. (2001), *The Internet Galaxy: Reflections on the Internet, Business, and Society*. Oxford: OUP.

Cavalieri, P. and Singer, P. (1993), *The Great Ape Project: Equality Beyond Humanity.* London: Fourth Estate.

Caygill, H. (2000), Liturgies of Fear: Biotechnology and Culture, in B. Adam, U. Beck and J. Van Loon (Eds) *The Risk Society and Beyond.* London: Sage.

Central Office of Information (1998), *Our Information Age – the Government's Vision.* London: Central Office of Information.

Chang, K. (2005), Think Small: Making Tiny Bits Useful, *Le Monde /The New York Times Supplement,* 19 March.

Chakravarti, A. and Little, P. (2003), Nature, Nurture and Human Disease, *Nature,* 421: 412–414.

Channel 4 (2000), *Big Brother: More Than a Game Show,* Channel 4 Television.

Chataway, J. (2005), Introduction: Is It Possible to Create Pro-Poor Agriculture Related Biotechnology? Special Issue on Perspectives on Institutions, *Agricultural Biotechnologies and Development ,* 17 (5): 597–610, published online 23 June 2005.

Chataway, J., Tait, J. and Wield, D. (2004), Understanding Company R&D Strategies in Agro-biotech: Trajectories and Blindspots. *Research Policy,* 33 (6–7): 1041–1057.

Chataway, J., Smith, J. and Wield, D. (2005), Partnerships for Building Science and Technology Capacity in Africa: UK and Canada Perspectives. Paper presented at the conference on Building Science and Technology Capacity with African Partners: An Africa-Canada-UK Exploration, London, 31 January–1 February (papers available on www.Scidev.Net) (accessed 2001).

Cheetham, K.A. and Grubstein, P.S.H. (2003), Nanomaterials and Venture Capital, *Nanotoday,* December, page 1. Available at http://www.materialstoday.com/ nanotoday/cheetham.pdf on 3rd April 2006.

Cheney, J. (1994), Nature/Theory/Difference, in K. Warren (Ed.) *Ecological Feminism.* London: Routledge.

Ciborra, C. (1999), Notes on Improvisation and Time in Organizations, *Accounting, Management and Information Technologies,* 9: 2, 77–94.

Clark, G. (1956), *The Later Stuarts 1660–1714.* 2nd edition. Oxford: Oxford University Press.

Clark, N., Yoganand, B., and Hall A. (2002), New Science, Capacity Development and Institutional Change: The Case of the Andhra Pradesh-Netherlands Biotechnology Programme (APNLBP), *International Journal of Technology Management and Sustainable Development,* 1 (3): 196–212.

Clement, A (1991), Designing Without Designers: More Hidden Skill in Office Computerization? in I. Eriksson, B. Kitchenham and K. Tijdens (Eds) *Women, Work and Computerization: Understanding and Overcoming Bias in Work and Education.* pp. 15–32. Amsterdam: North Holland.

Cohen, J. (2005), Poorer Nations Turn to Publicly Developed GM Crops, *Nature Biotechnology,* 23 (1): 27–33.

Cohen, J. and Paarlberg, R. (2004), Unlocking Crop Biotechnology in Developing Countries. A Report from the Field, *World Development,* 32 (9): 1563–1577.

Coleman, R. and Sim, J. (2000), You'll Never Walk Alone: CCTV Surveillance, Order and Neo-Liberal Rule In Liverpool City Centre, *British Journal of Sociology,* 51 (4): 623–639.

Collins, F.S., Patrinos, A., Jordan, E, Chakravarti, A, Gesteland, R and Walters, L.(1998), New Goals for the Human Genome Project, *Science,* 282 (5389): 682–689.

Commission of the European Communities (2002), *Life Sciences and Biotechnology – A Strategy for Europe*, Brussels, 23/1/2002om Luxembourg: Office for Official Publications of the European Communities.

Commission on Intellectual Property Rights (CIPR) (2002), *Integrating Intellectual Property Rights and Development Policy*, Report of the Commission on Intellectual Property Rights (CIPR), London.

Conway, G. (1997), *The Doubly Green Revolution: Food for all in the 21st Century*. Cornell: Cornell University Press.

Cooper, W. (2005), Raising Interactive Television Standards, *Cable & Satellite International*, July–August 2005, Available at http://informitv.com/opinion/2005/07/raisinginteractivetelevision/

Corea, G. (1985), *The Mother Machine*. London: The Women's Press.

Corner, J. (2000), What Can We Say about 'Documentary', *Media, Culture and Society*, 22: 681–688.

Cornford, J. (2000), The Virtual University Is...The University Made Concrete?, *Information, Communication & Society*, 3 (4): 508–525.

Cornia, A. (Ed.) (2004), *Inequality, Growth and Poverty in an Era of Liberalization and Globalization*. Oxford: Oxford University Press.

Cornish, W.R. (1989), *IP: Patents, Copyright, Trade Marks and Allied Rights*. London: Sweet & Maxwell.

Crichton, M. (2003), *Prey*. Avon: Harper Collins.

CSE: Microelectronics Group (1980), Microelectronics, *Capitalist Technology and the Working Class*. London: Conference of Socialist Economists.

Cudworth, E. (2005), *Developing Ecofeminist Theory: The Complexity of Difference*. Basingstoke: Palgrave.

Curran, J. and Seaton, J. (2003), *Power Without Responsibility*. London: Routledge.

Daar, A., Thorsteindöttir, H., Martin D.K., Smith, A.C., Nast, S. and Singer P. (2002), Top 10 Biotechnologies for Improving Health in Developing Countries, *Nature Genetics*, 32: 229–232.

Dahlbom, D. and Mathiassen, L. (1993), *Computers in Context: The Philosophy and Practice of Systems Design*. Oxford: NCC Blackwell Ltd.

Davenport, T. (2000), Putting the Enterprise into the Enterprise System, *Harvard Business Review*, 76 (4): 121–131.

David, P. (1990), The Dynamo and the Computer: An Historical Perspective on the Modern Productivity Paradox, *American Economic Review*, 80 (2): 355–361.

Davies, S. (1998), CCTV a New Battleground for Privacy, in C. Norris, J. Moran and G. Armstrong (Eds) *Surveillance, Closed Circuit Television and Social Control*, chapter 14. Aldershot: Ashgate Publishing Ltd.

Davis, M. (2007), Identity, Expertise and HIV Risk in a Case Study of Reflexivity and Medical Technologies, *Sociology*, 41 (6), 1003–1019 in press.

Davis, M., Hart, G., Imrie, J., Davidson, O., Williams, I. and Stephenson, J. (2002), HIV Is HIV to Me: Meanings of Treatments, Viral Load and Reinfection among Gay Men with HIV, *Health, Risk and Society*, 4 (1): 31–43.

Davis, M., Frankis, J. and Flowers, P. (2006), Uncertainty and Technological Horizon in Qualitative Interviews about HIV Treatment, *Health: An Interdisciplinary Journal for the Social Study of Health, Illness and Medicine*, 10 (3): 323–344.

DeLamarter, R. (1986), *Big Blue: IBM's Use and Abuse of Power*. New York: Dodd, Mead and Company.

Deleuze, G. and Guattari, F. (1987), *A Thousand Plateaus: Capitalism and Schizophrenia*, Trans. B. Massumi. London: Athlone Press.

Delia, J.G. (1970), The Logic Fallacy, Cognitive Theory, and the Enthymeme: A Search for the Foundations of Reasoned Discourse, *Quaterly Journal of Speech*, 56: 140–148.

Department of Health (2003), *Our Inheritance, Our Future, Realising the Potential of Genetics in the NHS*. Norwich: The Stationery Office.

Department of Trade and Industry (1999), *Biotechnology Clusters*, Department of Trade and Industry, London: HMSO.

Department of Trade and Industry (2001), *Science and Innovation Strategy*, Department of Trade and Industry, London: HMSO.

Department of Trade and Industry (2003), *GM Nation? The Findings of the Public Debate*, Department of Trade and Industry, London: HMSO.

Dickens, P. (1996), *Reconstructing Nature: Alienation, Emancipation and the Division of Labour*. London: Routledge.

Dickens, P. (2001), Linking the Social and Natural Sciences: Is Capital Modifying Human Biology in Its Own Image? *Sociology*, 35 (1): 93–110.

Ditton, J., (1998), Public Support for Town Centre CCTV Schemes: Myth or Reality, in C. Norris, J. Moran and G. Armstrong (Eds.) *Surveillance, Closed Circuit Television and Social Control*, chapter 12. Aldershot: Ashgate Publishing Ltd.

Ditton, J., Short, E., Phillips, S. Norris, C and Armstrong, G. (1999), *The Effect of Closed Circuit Television on Recorded Crime Rates and Public Concern About Crime in Glasgow*. Edinburgh: The Scottish Office Central Research Unit.

Dohse, D. (2000), Technology Policy and the Regions – The Case of the BioRegio Contest, *Research Policy*, 20 (9): 1111–1133.

Dorfman, R. (2002), The Future of Reality Television, *Exquisite Corpse: A Journal of Letters and Life*, Available at www.corpse.org/issue_11/critiques/dorfman.h

Douglas, E. (1999), The Guardian Profile Robert May: Testing, Testing, *The Guardian*, 30 October 1999.

Douthwaite, B. (2002), *Enabling Innovation: A Practical Guide to Understanding and Fostering Technological Change*. London: Zed Books.

Dovey, J. (1996), *Fractal Dreams: New Media in Social Context*. London: Lawrence and Wishart.

Dovey, J. (2000), *Freakshow: First Person Media and Factual Television*. London: Pluto.

Dowsett, G. and McInnes, D. (1996), Gay Community, AIDS Agencies and the HIV Epidemic in Adelaide: Theorising 'Post AIDS', *Social Alternatives*, 15 (4): 29–32.

Drake, R. (2005), The New Genomics and the New Nutrition. Paper Presented at the ESRC INNOGEN Research Centre International Conference on Evolution in the Life Sciences, Edinburgh, 23–25 February 2005.

Drews, J. (2003), Strategic Trends in the Drug Industry, *Drug Disc Today*, 8, 2003: p. 411.

Duncan, M. (1981), *Microelectronics: Five Areas of Subordination in Science, Technology and the Labour Process*, London: CSE Books.

Dutfield, G. (2004), *Intellectual Property, Biogenetic Resources and Traditional Knowledge: A Guide to the Issues*. London: Earthscan.

Eder, K. (1996), *The Social Construction of Nature*. London: Sage.

Edwards, J., -Hirsch, E. and Price, F. (1999), *Technologies of Procreation: Kinship in the Age of Assisted Conception*, 2nd edition. London: Routledge.

Enriquez, J. and Goldberg, R. A. (2000), Transforming Life, Transforming Business: The Life-Science Revolution. March, *Harvard Business Review*, 78 (2): 94–10.

European Commission (1993), *Growth, Competitiveness and Employment*. COM(93) 700, 5 December.

European Commission (1995), *Conclusions of G7 Summit Information Society Conference*. DOC/95/2 of 1995–02-26. Available at http://europa.eu.int/ISPO/docs/services/docs/1997/doc_95_2_en.pdf, 18 July 2007.

European Council (1994), *Europe and the Global Information Society*. Brussels: European Council.

Evans, C. (1979), *The Mighty Micro*, London: Gollancz.

Ezrahi, Y. (1998), *Rubber Bullets: Power and Conscience in Modern Israel*. University of California Press.

Ezzy, D. (2000), Illness Narratives: Time, Hope and HIV, *Social Science and Medicine*, 50: 605–617.

FDA (2004), *Innovation or Stagnation, Challenge and Opportunity on the Critical Path to New Medicinal Products*. U.S. Department of Health and Human Services, Food and Drug Administration.

Farrel, S. (2001), *Gringrich, Toffler and Gore: A peculiar trio: Democrats in Drag: Third Way Fall From Grace III*. Blog posted July 9 2001. Http://www.enterstageright.com/archive/articles/0701thirdwayp3.htm (accessed on 25 May 2008).

Feather, J. (1988), *A History of Publishing*. London: Croom Helm.

Featherstone, M. (1991), *Consumer Culture and Postmodernism*. London: Sage.

Febvre, L. and Martin, H-J. (1976), *The Coming of the Book*. London: Verso.

Feigenbaum, R. and McCorduck, P. (1984), *The Fifth Generation: Artificial Intelligence and Japan's Computer Challenge to the World*. London: Michael Joseph.

Fetveit, A. (1999), Reality TV in the Digital Era: A Paradox in Visual Culture, *Media, Culture and Society*, 21: 787–804.

Firestone, S. (1988), *The Dialectic of Sex*. London: The Women's Press.

Fisher, W.R., (1985a), The Narrative Paradigm: In the Beginning, *Journal of Communication*, 35: 74–89.

Fisher, W.R. (1985b), The Narrative Paradigm: An Elaboration, *Communication Monographs*, 52: 347–367.

Fisher, W.R. (1989), Clarifying the Narrative Paradigm, *Communication Monographs*, 56: 55–58.

Fiske, J. (1990), *Introduction to Communication Studies*, 2nd Edition. London: Routledge.

Flowers, P., Knussen, C. and Duncan, B. (2001), Re-appraising HIV Testing among Scottish Gay Men: The Impact of New HIV Treatments, *Journal of Health Psychology*, 6 (6): 665–678.

Flowers, P., Rosengarten, M., Davis, M., Hart, G. and Imrie, J. (2004), *The Experiences of HIV Positive Black Africans Living in the UK*, Working paper, Glasgow, Transitions in HIV Project, Glasgow Caledonian University.

Flowers, P., Davis, M., Hart, G., Imrie, J., Rosengarten, M. and Frankis, J. (2006), Diagnosis and Stigma and Identity amongst HIV Positive Black Africans Living in the UK, *Psychology and Health*, 21 (1): 109–122.

Forester, T. (1980), *The Microelectronics Revolution: The complete Guide to the New Technology and its Impact on Society*. Oxford: Basil Blackwell.

Foucault, M. (1979), *Discipline and Punish: The Birth of the Prison*, Trans. A. Sheridan, Harmondsworth: Penguin.

Franklin, A. (1999), *Animals and Modern Cultures: A Sociology of Human-Animal Relations in Modernity*. London: Sage.

Franklin, S. (1997), *Embodied Progress: A Cultural Account of Assisted Conception*. London: Routledge.

Freeman C. and Louca F. (2002), *As Time Goes By: From the Industrial Revolution to the Information Revolution*. Oxford: Oxford University Press.

Friedman, A. (1989), *Computer Systems Development*, London: J. Wiley and Sons.

Fussey, P. (2004), New Labour and New Surveillance: Theoretical and Political Ramifications of CCTV Implementation in the UK, *Surveillance and Society CCTV Special Edition* (Eds C. Norris, M. McCahill, and D. Wood), 2 (2/3): 251–269. Available at http://www.surveillance-and-society.org/articles2(2)/newlabour.pdf

Galliers, R.D. (2004), Reflections on Information Systems Strategizing, in C. Avgerou, C. Ciborra and F. Land (Eds) *The Social Study of Information and Communication Technology: Innovation, Actors, and Contexts*, pp. 231–262. Oxford: Oxford University Press.

Galvin, R. (2002), Disturbing Notions of Chronic Illness and Individual Responsibility: Towards a Genealogy of Morals, *Health: An Interdisciplinary Journal for the Social Study of Health, Illness and Medicine*, 6 (2): 107–137.

Garnham, N. (1990), *Capitalism and Communication: Global Culture and the Economics of Information*. London: Sage Publications.

Gatens, M. (1996), *Imaginary Bodies: Ethics, Power and Corporeality*. London: Routledge.

Gates, B. (2000), *Economics and the Digital Divide*. Comments to White House panel discussion hosted by President Clinton on 5 April 2000. Available at http://www.microsoft.com/presspass/features/2000/04–05wh.mspx

Gibbons, M., Limoges, C., Nowotny, H., Scott, P. Schwartman, S. and Trow, M. (1994), *The New Production of Knowledge: The Dynamics of Science and Research in Contemporary Societies*. London: Sage.

Giddens, A. (1991), *Modernity and Self Identity. Self and Society in the Late Modern Age*. Cambridge: Polity Press.

Giddens, A. (1977), *Studies in Social and Political Theory*. New York: Basic Books.

Giddens, A. (1998), *The Third Way: The Renewal of Social Democracy*. Cambridge: Polity.

Giesecke, S. (2000), The Contrasting Roles of Government in the Development of Biotechnology Industry in the US and Germany, *Research Policy*, 29 (2–3): 205–223.

Gill, M. and Spriggs, A. (February, 2005), Assessing the Impact of CCTV, Home Office Research Study 292, Home Office Research, Development and Research Directorate. Available at http://www.popcenter.org/Responses/Supplemental_Material/video_surveillance/Gill&Spriggs_2005.pdf

Gore, A. (1994a), Address to the International Telecommunications Union, 21 March. Available at http://www.interesting-people.org/archives/interesting-people/199403/msg00112.html (accessed on 18 July 2007).

Gore, A. (1994b), Remarks by Vice President Al Gore to the International Telecommunication Union Plenipotentiary Conference, Kyoto, Japan, 22 September. Available at http://www.friends-partners.org/oldfriends/telecomm/al.gore.speech (accesssed on 18 July 2007).

Grade, M. (2004), Opening Speech Given to the Interactive TV Show Europe 2004 in Barcelona, 14 October 2004. Available at www.bbc.co.uk/pressoffice/speeches/stories/grade_barcelona.shtm

Graham, S. (1998), Toward the Fifth Utility? On the Extension and Normalisation of Public CCTV, in C. Norris, J. Moran, and G. Armstrong (Eds.) *Surveillance, Closed Circuit Television and Social Control*. Aldershot: Ashgate Publishing.

Grant, D., Hall, R., Wailes, N. and Wright, C. (2006), The False Promise of Technological Determinism: The Case of Enterprise Resource Planning Systems, *New Technology, Work and Employment*, 21 (1): 1–15.

Green, G. and Smith, R. (2004), The Psychosocial and Health Care Needs of HIV-Positive People in the United Kingdom following HAART: A Review, *HIV Medicine*, 5, Supplement 1, 1–46.

Greene, R. and Ward, D. (2002), Let's Call it HIV Infection, not AIDS. Poster presented at the 14th International AIDS Conference, Barcelona, 7–14 July 2002.

Greer, G. (1984), *Sex and Destiny: The Politics of Human Fertility*. London: Pan Books.

Greer, G. (1999), *The Whole Woman*. London: Doubleday.

Grosz, E. (1994), *Volatile Bodies: Towards a Corporeal Feminism*. London: Allen and Unwin.

Haigh, T. (2001), The Chromium-Plated Tabulator: Institutionalizing an Electronic Revolution, 1954–58, in *IEEE Annals of the History of Computing*, October–December 2001, 23 (4), Los Alamitos, CA: IEEE Computer Society.

Halberstram, J. (1991), Automating Gender: Postmodern Feminism in the Age of the Intelligent Machine, *Feminist Studies*, 3: 439–460.

Halberstram, J. and Livingston, I. (1995), *Posthuman Bodies*. Bloomington: Indiana University Press.

Hall, A., Yoganand, B., Crouch, J. and Clark N.G. (2004), The Evolving Culture of Science in the Consultative Group on International Agricultural Research (CGIAR): Concepts for Building a New Architecture of Innovation in Agri-Biotechnology, in A.J. Hall, B. Yoganand, R.V. Sulaiman, Rajeswari S. Raina, C. Shambu Prasad, Guru C. Naik and N.G. Clark (Eds) *Innovations in Innovation*, pp. 135–162. Pantancheru, India: ICRISAT.

Haraway, D. (1991), *Simians, Cyborgs and Women: The Reinvention of Nature*. London: Free Association Press.

Haraway, D. (1997), *Modest_Witness@Second_Millennium. FemaleMan_Meets_OncoMouse*. London: Routledge.

Haraway, D. (2003), *The Companion Species Manifesto: Dogs, People and Significant Otherness*. Chicago: Prickly Paradigm Press.

Hayles, N.K. (1999), *How We Became Posthuman: Virtual Bodies in Cybernetics, Literature and Informatics*. Chicago: the University of Chicago Press.

Headrick, D.R. (1981), *The Tools of Empire: Technology and European Imperialism in the 19th Century*. Oxford: Oxford University Press.

Headrick, D.R. (2000), *When Information Came of Age: Technologies of Knowledge in the Age of Reason and Revolution, 1700–1850*. Oxford: Oxford University Press.

Health Protection Agency (2004), *AIDS/HIV Quarterly Surveillance Tables: Cumulative UK Data to End June 2004*. London: Health Protection Agency.

Heap, N., Thomas, R., Einon, G., Manson, R. and Mackay, H. (1995), *Information Technology and Society: A Reader*. London: Sage.

Hempel, L. and Töpfer, E. (August 2004), *CCTV in Europe Final Report*, Working Paper No. 15, *Urban Eye*. Available at http://www.urbaneye.net/results/ue_wp15.pdf

Hill, A. (2005), *Reality TV: Audiences and Popular Factual Television*. London: Routledge.

Hill, C.P. (1965), *Who's Who in History, Volume III, England 1603–1714*. Oxford: Basil Blackwell.

Hirschheim, R.A. (1986), The Effect of A Priori Views on the Social Implications of Computing: The Case of Office Automation, *Computing Surveys*, 18 (2): 165–195.

Hirschheim, R. and Newman, M. (1991), Symbolism and Infomration Systems Development: Myth, Metaphor and Magic, *Information Systems Research* 2 (1): 29–62.

Hobsbawm, E. and Ranger, T. (1983), *The Invention of Tradition*. Cambridge: Cambridge University Press.

Horstkotte-Wesseler, G. and Byerlee, D. (2000), Agricultural Biotechnology and the Poor: The Role of Development Assistance Agencies, in M. Qaim, A.F. Krattiger, J. von Braun (Eds) *Agricultural Biotechnology in Developing Countries: Towards Optimizing the Benefits for the Poor*. Kluwer Academic Publishers: Dordrecht: Paı́ses Bajos.

Horrobin, D.F. (2001), Realism in Drug Discovery – Could Cassandra be Right?, *Nature Biotechnology*, 19: 1099–1100.

Horrobin, D.F. (2003), Modern Biomedical Research: An Internally Self-Consistent Universe with Little Contact with Medical Reality?, *Nature Reviews Drug Discovery*, 2: 151–154.

House of Commons Science and Technology Committee (1995), *Human Genetics: The Science and its Consequences*, London: HMSO.

Hughes, P. (1990), Today's Television, Tomorrow's World, in A. Goodwin and G. Whannel (Eds) *Understanding Television*. London: Routledge.

Humm, P. (1998), Real TV: Camcorders, Access and Authenticity, in C.Geraghty and D. Lusted (Eds) *The Television Studies Book*. London: Arnold.

Ifrah, G. (2000), *The Computer and the Information Revolution. The Universal History of Numbers*, vol III, Trans. E.F. Harding. London: Harvill Press.

Infotech (1971), *Infotech State of the Art Report 1: The Fourth Generation*. Maidenhead: Infotech.

Ingold, T. (2000), *The Perception of the Environment: Essays on Livelihood, Dwelling and Skill*. London: Routledge.

International-Collaboration-on-HIV-Optimism (2003), HIV Treatments Optimism among Gay men: An International Perspective, *Journal of Acquired Immune Defiency Syndromes*, 32, 545–550.

IITF (1993), *The National Information Infrastructure: Agenda for Action – Executive Summary*. Available at http://www.ibiblio.org/nii/NII-Executive-Summary.html (accessed on 18 July 2007).

Jamieson, K.H. (1985), The idea of a University in an electronic age, in, *Precis*, the faculty/ staff newsletter of the University of Maryland, College Park Campus, 16 (10).

Jasanoff, S. (2005), *Designs on Nature: Science and Democracy in Europe and the United States*. Princeton University Press.

Jung, C.G. (1968), *Psychology and Alchemy*. Collected Works, vol 12. Princeton: Princeton University Press.

Kahane, B. and Reitter, R. (2002), Narrative Identity: Navigating between 'Reality' and 'Fiction', in B. Moingeon and G. Soenen (Eds) *Corporate and Organization Culture*. London: Routledge.

Kallinikos, J. (2004), Deconstructing Information Packages: Organizational and Behavioural Implications of ERP Systems, *Information Technology & People* 17 (1): 8–30.

Karembu, M. (2004), Biotechnology Transfer Strategies: Some Lessons for Policy in Taking Products to Farmers in East Africa. Paper Presented at ESRC Science in Society/INNOGEN Workshop on Technology-Based PPPs and Innovation Systems in African Agriculture, (19 November 2004).

Kauffman, S. (1995), *At Home in the Universe: The Search for Laws of Self-Organization and Complexity*. Oxford: Oxford University Press.

Keeley, J. and Scoones, I. (2003), Understanding Environmental Policy Processes, Cases from Africa. London: Earthscan.

Keller, E.F. (1995), *Refiguring Life: Metaphors of Twentieth-Century Biology*. New York: University of Colombia Press.

Keller, E.F. (2000), *Century of the Gene*. Cambridge, MA: Harvard University Press.

Kilborn, R. and Izod, J. (1997), *An Introduction to Television Documentary: Confronting Reality*. Manchester: Manchester University Press.

Kilborn, R. (1998), Shaping the Real: Democratisation and Commodification in UK Factual Broadcasting, *European Journal of Communication*, 13 (2): 201–218.

Kimbrell, A. (1993), *The Human Body Shop: the Engineering and Marketing of Life*, London: Harper Collins.

Kling, R. and Iacono, S. (1984), The Control of Information Systems Developments after Implementation, *Communications of the ACM*, 27 (12): 1218–1226.

Krattiger, A. (2002), *Public–Private Partnerships for Efficient Proprietary Biotech Management and Transfer and Increased Private Sector Investments*. A Briefings Paper with Six Proposals Commissioned by UNIDO. IP Strategy Today No. 4.

Krugman, P. (2002), For Richer, *New York Times Magazine*, 20 October 2002.

Kubinyi, H. (2003), Drug Research: Myth Hype and Reality, *Nature Rev Drug Disc.* 2, 2003, 665–668.

Kumar, K. and van Hillegersberg, J. (2000), ERP Experiences and Evolution, *Communications of the ACM*, 43 (4): 23–26.

Latour, B. (1993), *We Have Never Been Modern*. Hemel Hempstead: Harvester Wheatsheaf.

Latour, B. (2004, Winter), Why has Critique Run Out of Steam? From Matters of Fact to Matters of Concern, *Critical Inquiry*, 30 (20): 225–248.

Le Hir, P. and Cabret, N. (2005), Des activistes grenoblois contre les 'nécrotechnologies', *Le Monde*, 17 June 2005.

Lessig, L. (2002) *The Future of Ideas: The Fate of the Commons in a Connected World*. New York: Random House.

Lenaghan, J. (1998), *Brave New NHS?* London: IPPR.

Levi-Strauss, C. (1966), *The Savage Mind*. Chicago: Chicago University Press.

Levy, M. and Gunter, B. (1988), *Home Video and the Changing Nature of the Television Audience*. London: John Libbey.

Lewontin, R. (1993), *Biology as Ideology: The Doctrine of DNA*. New York: Harper Perennial.

Lindpaintner, K. (2002a), The Impact of Pharmacogenetics and Pharmacogenomics on Drug Discovery, *Nature Reviews Drug Discovery*, 1: 463–469.

Lindpaintner, K. (2002b), Pharmacogenetics and the Future of Medical Practice. *British Journal of Clinical Pharmacology*, 54 (2): 221–230.

Lukacs, G. (1971), *History and Class Consciousness*. London: Merlin.

Lykke, N. (1996), Between Monsters, Goddesses and Cyborgs: Feminist Confrontations with Science, in N. Lykke and R. Braidotti (Eds) *Between Monsters, Goddesses and Cyborgs: Feminist Confrontations with Science, Medicine and Cyberspace*. London: Zed.

Lyon, D. (1988), *The Information Society: Issues and Illusions*. Cambridge: Polity Press.

Lyon, D. (2001), *Surveillance Society, Monitoring Everyday Life*. Buckingham: Open University Press.

Mabert, V., Soni, A. and Venkataramanan, M. (2001), Enterprise Resource Planning: Common Myths Versus Evolving Reality, *Business Horizons*, May–June: 69–76.

McChesney, R. (1999), *Rich Media, Poor Democracy: Communication Politics in Dubious Times*. New York: The New Press.

McClean, P. (1997), *Historical Events in the DNA Debate*. Available at http://www.ndsu. nodak.edu/instruct/mcclean/plsc431/debate/debate3.htm (accessed on 26 March 2005).

Macdonald, M. (2007), Television Debate, Interactivity and Public Opinion: The Case of the BBC's 'Asylum day', *Media, Culture and Society*, 29: 679.

Macintyre, R. (1999), *Mortal Men: Living with Asymptomatic HIV*. New Brunswick: Rutgers University Press.

Mackinnon, F. (1985), Notes on the History of English Copyright, in M. Drabble (Ed.) *The Oxford Companion to English Literature*. Oxford: Oxford University Press.

Mackintosh, I. (1986), *Sunrise Europe*, Oxford: Basil Blackwell.

McLaughlin, J., Rosen, P., Skinner, D. and Webster, A. (1999), *Valuing Technology: Organisations, Culture and Change*. London: Routledge.

McLuhan, M. (1964), *Understanding Media: The Extensions of Man*, New York: McGraw-Hill.

Macnaghten, P. and Urry, J. (1998), *Contested Natures*. London: Sage.

Mahdi, S. (2004), The Pharmaceutical Industry, Unpublished Biblio-metric Dataset SPRU. University of Sussex, UK.

March, J.G. and Simon, H.W. (1958), *Organizations*. New York: Wiley.

Marcuse, H. [1955] (1998), *Eros and Civilization*. London: Routledge.

Marcuse, H. [1964] (2002), *One Dimension Man*. London: Routledge.

Margulis, L. and Sagan, D. (1986), *Microcosmos*. New York: Summit.

Martin, J. (1978), *The Wired Society: A Challenge for Tomorrow*. New Jersey: Prentice-Hall.

Martin, J. and Powers, M.E. (1983), Truth or Corporate Propaganda: The Value of a Good War Story, in L.R. Pondy, P.J. Frost, G. Morgan and T.C. Dandridge (Eds) *Organizational Symbolysm*, pp. 93–107, Greenwich, CT: JAI Press.

Marx, K. (2002), *The Communist Manifesto*. London: Penguin.

Mason, R.O. (1969), A Dialectical Approach to Strategic Planning, *Management Science*, 15: 403–414.

Mattelart, A. (2000), Archéologie de la 'Société de l'Information, *Le Monde Diplomatique*, August.

Merchant, C. (1980), *The Death of Nature: Women, Ecology and the Scientific Revolution*. San Francisco: Harper and Row.

Merle, R. (1983), *Malevil*. Paris: Editions Gallimard.

Midgley, M. (1996), *Utopias, Dolphins and Computers: Problems of Philosophical Plumbing*. London: Routledge.

Milburn, A. (2001), Speech by the Secretary of State for Health at the Institute of Human Genetics, International Centre for Life in Newcastle-upon-Tyne (19 April 2001). Available at: http://www.dh.gov.uk/en/News/Speeches/Speecheslist/DH_4000758

Millennium Project (2004), Interim Report of Task Force 10 on Science, Technology and Innovation. Commissioned by the UN Secretary General and the UN Development Group, New York.

Molière, J-B.P. (2005), *Le bourgeois gentilhomme*. Paris: Hachette Education.

Moravec, H. (1988), *Mind Children: the Future of Robot and Human Intelligence* Cambridge, MA: Harvard University Press.

Moschella, D.C. (1997), *Waves of Power, The Dynamics of Global Technology Leadership*. New York: Amacom.

Mosco, V. (2005), *The Digital Sublime: Myth, Power and Cyberspace*. Cambridge MA: The MIT Press.

Mumford, L. (1967), *The Myth of the Machine: Technics and Human Development.* London: Secker & Warburg Ltd.

Naess, A. (1979), Self-realization in Mixed Communities of Humans, Bears, Sheep and Wolves, *Inquiry*, 16: 95–100.

NAM (2002a), Haart and HIV Transmission, *AIDS Treatment Update*, 118: 1–5. NAM (2002b), NAM Factsheet 11: Viral Load, *National AIDS Manual*, London: NAM Publications. NAM (2003), Types of Side Effects, www.aidsmap.com.

Nature Editorial (2003), Facing Our Demons, *Nature Reviews Drug Discovery*, 2 (2): 87.

Nichols, B. (1999), Reality TV and Social Perversion, in P.Marris and S.Thornham (Eds) *Media Studies: A Reader*, 2nd edition. Edinburgh: Edinburgh University Press.

Nightingale, P. (2000), Economies of Scale in Experimentation: Knowledge and Technology in Pharmaceutical R&D, *Industrial and Corporate Change*, 9 (2): 315–359.

Noble, D.F. (2001/2004), *Digital Diploma Mills: The Automation of Higher Education*, New York: Monthly Review Press; /Delhi: Aakar Books.

Norris, C., McCahill, M. and Wood, D. (2004), Editorial. The Growth of CCTV: A Global Perspective on the International Diffusion of Video Surveillance in Publicly Accessible Space, *Surveillance and Society CCTV Special Edition* (Eds C. Norris, M. McCahill, and D. Wood 2 (2/3): 110–135. Available at http://www.surveillance-and-society.org/articles2(2)/editorial.pdf

Norris, C. and Armstrong, G. (1999), *The Rise of the Maximum Surveillance Society.* Oxford: Berg.

Oakley, A. (2002), *Gender on Planet Earth.* Cambridge: Polity.

OECD (1997), *Biotechnology and Medical Innovation: Socio-economic Assessment of the Technology, the Potential and the Products.* Paris: OECD.

OECD (1998), *Economic Aspects of Biotechnologies Related to Human Health Part II: Biotechnology, Medical Innovation and the Economy: The Key Relationships.* Paris: OECD.

OECD, Committee for Scientific and Technological Policy at Ministerial Level (2004), Science, Technology and Innovation for the 21st Century, Final Communique meeting, 29–30. Paris: OECD.

Officer, L.H. (2007), *Purchasing Power of British Pounds from 1264 to 2006.* MeasuringWorth. com, March 2007.

Oliver, D. and Romm, C. (2002), Justifying Enterprise Resource Planning Adoption, *Journal of Information Technology*, 17: 199–213.

ONS1 and ONS2: Office of National Statistics, *Family Spending* 2006 and 1999–2000 editions. Available at http://www.statistics.gov.uk/STATBASE

Orwell, G. (1950), *1984*, New York: Signet Classic.

Oyelaran-Oyeyinka, B. (2005), Partnerships for Building Science and Technology Capacity in Africa. Paper presented at the conference on Building Science and Technology Capacity with African Partners: An Africa-Canada-UK Exploration, London, 31 January–1 February. Available at www.Scidev.Net

Pollard, E, Barkworth, R, Sheppard, E, Tamkin, P. (2005) *Researching the Independent Production Sector: a Focus on Minority Ethnic Led Companies*, London: Pact/UK Film Council.

Parker, J. (2000), *Total Surveillance*, London: Piatkus.

Patel, R. (2007), *Stuffed and Starved: Markets, Power and the Hidden Battle for the World's Food System.* London: Portobello Books.

Patton, L.L. and Doniger, W. (1998), *Myth and Method.* Charlottesville and London: University Press of Virginia.

Pawson, R. and Tilley, N. (1994), What Works in Evaluation Research?, *British Journal of Criminology*, 34 (30): 291–306.

Perelman, M. (2002), *Steal This Idea: IP Rights and the Corporate Confiscation of Creativity*. New York and Basingstoke: Palgrave.

Pfaff, W. (2005), Traditional Culture Strikes Back, *International Herald Tribune*, 21 July 2005.

Pfeiffer, N. (1987), Artificial Insemination, In-Vitro Fertilisation and the Stigma of Infertility, in, M. Stanworth (ed.) *Reproductive Technologies*. Cambridge: Polity Press.

Pierret, J. (2001), Interviews and Biographical Time: The Case of Long-Term HIV Nonprogressors, *Sociology of Health and Illness*, 23 (2): 159–179.

Piketty, T. and Saez, E. (2003), Income Inequality in the United States, 1913–1998, *Journal of Political Economy*, 111 (5): 1004–1042, with data updated in November 2004.

Pratley, N. (2003), The Drugs Don't Work, *The Guardian*, 23 November.

Price, F. (1999), Solutions for Life and Growth? Collaborative Conceptions in Reproductive Medicine, in J. Edwards, S, Franklin, E. Hirsch, F. Price and M. Strathern (Eds.) *Technologies of Procreation: Kinship in the Age of Assisted Conception*, 2nd edition. London: Routledge.

Prieur, C. (2005), L'affaire de l'amiante révèle les carences du système de santé au travail, *Le Monde*, 11 September 2005.

Rabinow, P. (1992), Artificiality and Enlightenment, in J. Crary and S. Kwinter (Eds) *Incorporations*. New York: Zone Books.

Race, K. (2001), The Undetectable Crisis: Changing Technologies of Risk, *Sexualities*, 4 (2): 167–189.

Raymond, J. (1993), *Women as Wombs: Reproductive Technologies and the Battle over Women's Freedom*. San Francisco: Harper and Row.

Rayner, S. (2003), Democracy in the Age of Assessment: Reflections on the Roles of Expertise and Democracy in Public Sector Decision Making, *Science and Public Policy*, 30 (3): 163–171.

Reich, R.B. (1999), The Political Paradox of Inequality, paper presented at the *Inequality Summer Institute '99*. Available at www.ksg.harvard.edu/inequality/Summer/Summer99/privatepapers/Reich.PDF (accessed on 18 July 2007).

Reingold, H. (1991), *Virtual Reality*. New York: Summit Books.

Rhodes, T. and Simic, M. (2005), Transition and the HIV Risk Environment, *British Medical Journal*, 331: 220–223.

Rhodes, T., Bernays, S., Davis, M. and Green, J. (2006), *Experiencing HIV Treatment in the Context of Uncertainty: A Qualitative Study in Serbia/Montenegro*. ESRC Small Grants Scheme, London School of Hygiene and Tropical Medicine.

Robey, D., Ross, J. and Boudreau, M.C. (2002), Learning to Implement Enterprise Systems: An Exploratory Study of the Dialectics of Change, *Journal of Management Information Systems*, 19 (1): 17–46.

Robins, K. (1996), *Into the Image: Culture and Politics in the Field of Vision*. London: Routledge.

Rofes, E. (1998), *Dry Bones Breathe: Gay Men Creating Post-AIDS Identities and Cultures*. New York: Harrington Park Press.

Rosenberg, N. (1979), Technological Interdependence in the American Economy, *Technology and Culture*, 20: 25–51.

Rosengarten, M., Race, K. and Kippax, S. (2001), 'Touch Wood, Everything Will Be OK': *Gay Men's Understandings of Clinical Markers in Sexual Practice*, Sydney: National Centre in HIV Social Research.

Rosengarten, M., Imrie, J., Flowers, P., Davis, M. and Hart, G. (2004), After the Euphoria: HIV Medical Technologies from the Perspective of Their Prescribers, *Sociology of Health and Illness*, 26 (5): 575–596.

Sampedro, V. (2001), *New Genres in Commercial Television and their Effect on Public Opinion*. Expert Seminar on The European Convention on Transfrontier Television in an Evolving Broadcasting Environment, European Convention on Transfrontier Television, Strasbourg, 6 December 2001. Available at www.coe.int/T/e/Human_Rights/Media/2_Transfrontier_Television/Texts_and_documents/T-TT(2001)er4_en.asp

Samuel, R. and Thompson, P. (1990), *The Myths We Live By*. London: Routledge.

Samuels, E. (2002), *The Illustrated History of Copyright*. New York: Thomas Dunne Books.

Sassen, S. (2002), Towards a Sociology of Information, *Current Sociology*, 50 (3): 365–388.

Sawyer, S (2000), Packaged Software: Implications of the Differences from Custom Approaches to Software Development, *European Journal of Information Systems*, 9: 47–58.

Schumpeter, J. (1989), *Essays: On Entrepreneurs, Innovations, Business Cycles and the Evolution of Capitalism*, Edison, NJ: Transaction Publishers.

Scott, S. and Wagner, E. (2003), Networks, Negotiations, and New Times: The Implementation of Enterprise Resource Planning into an Academic Administration, *Information and Organization*, 13: 285–313.

Segal, R.A. (1998), Does Myth have a Future?, in L.L. Patton and W. Doniger (Eds.) *Myth and Method*. Charlottesville and London: University Press of Virginia.

Senker, J., Enzing, C., Joly, P. and Reiss, T. (2000), European Exploitation of Biotechnology – do government policies help, *Nature Biotechnology*, 18: 605–609.

Senker, P., (2000), A Dynamic Perspective on Technology, Economic Inequality and Development, in S. Wyatt, F. Henwood, N. Miller, and P. Senker (Eds) *Technology and In/equality: Questioning the Information Society*, pp. 197–217. London: Routledge.

Sharpe, R. (2004), Health On-line Services, in W. Abbott, N. Blankley, J. Bryant and S. Bullas (Eds.) *Information in Healthcare*. Swindon, UK: British Computer Society, Healthcare Informatics Committee.

Shattuc, J. (1998), 'Go Rikki': Politics, Perversion and Pleasure in the 1990s, in C. Geraghty and D. Lusted (Eds) *The Television Studies Book*. London: Arnold.

Shaviro, S. (1995), Two Lessons from Burroughs, in J. Halberstram and I. Livingston (Eds.) *Posthuman Bodies*. Bloomington: Indiana University Press.

Shelley, M. (2004), *Frankenstein*. Pocket Books.

Shilling, C. (2003), *The Body and Social Theory*, 2nd edition. London: Sage.

Shiva, V. (1988), *Staying Alive: Women, Ecology and Development*. London: Zed.

Shiva, V. (1993), *Monocultures of the Mind*. London: Zed.

Shiva, V. (1998), *Biopiracy: The Plunder of Nature and Knowledge*. Dartington: Green Books.

Silverstein, K. (1999), Buck Rogers Rides Again, *The Nation*, 25 October.

Singer, P. (1979), *Animal Liberation*. New York: Avon Books.

Skinns, D. (1998), Crime Reduction, Diffusion and Displacement: Evaluating the Effectiveness of CCTV, in C. Norris, J. Moran, and G. Armstrong (Eds) *Surveillance, Closed Circuit Television and Social Control*. Aldershot: Ashgate.

Smith J. (2004), The Anti-Politics Gene: Biotechnology, Ideology and Innovation Systems in Kenya. INNOGEN Working Paper. Available at www.innogen.ac.uk (accessed April 2005).

Soh, C., Kien, S.S. and Tay-Yap, J. (2000), Cultural Fits and Misfits: Is ERP a Universal Solution? *Communcations of the ACM*, 43 (4): 47–51.

Sontag, S. (1988), *AIDS and its Metaphors*. London, Penguin.

Soper, K. (1995), *What is Nature?* Oxford: Blackwell.

Spielman, D. and von Grebmer, K. (2004), *Public–Private Partnerships in Agricultural Research: An Analysis of Challenges Facing Industry and the Consultative Group on International Agricultural Research*. Environment and Production Technology Division Working Paper No. 113. Washington DC.: International Food Policy Research Institute (IFPRI).

Stabile, C.A. (1994), *Feminism and the Technological Fix*. Manchester: Manchester University Press.

Stallman, R. (1998), The GNU Operating System and the Free Software Movement, in C. DiBona, S. Ockman, and S. Stone (Eds) *Opensources: Voices from the Open Software Revolution*. California: O'Reilly and Associates.

Steelside Solutions Limited (May 2003), *Future Opportunities for Telephony and Interactive Services Provision to UK Broadcasters*. Available at http://www.e-consultancy.com/publications/tv_interactivity/

Stepulevage, L. (2003), Computer-Based Office Work: Stories of Gender, Design, and Use, in *IEEE Annals of the History of Computing*, 25 (4): 67–72.

Stepulevage, L. and Mukasa, M. (2005), Implementation of Large Scale Software Applications: Possibilities for End-User Participation, in J. Archibald, J. Emms, F. Grundy, J. Payne, and E. Turner (Eds.) *The Gender Politics of ICT*. London: Middlesex University Press.

Sterling, B. (1992), *The Hacker Crackdown: Law and Disorder on the Electronic Frontier*. Harmondsworth: Penguin.

Stoker B. (2003), *Dracula*. London: Penguin Books.

Stone, A.R. (1995), *The War of Desire and Technology at the Close of the Mechanical Age*. Cambridge, MA: MIT Press.

Stratton, J. and Ang, I. (1994), Sylvania Waters and the Spectacular Exploding Family, *Screen*, 35: 1.

Suchman, L. (2002), Located Accountabilities in Technology Production, in *Scandinavian Journal of Information Systems*, 14 (2): 91–105.

Swiss, Re. (2004), *Nanotechnology – Small Matter, Many Unknowns*. Swiss Re report.

Tarkovsky, A. (2002), *Stalker*. Image Entertainment [DVD].

Taverne, D. (2005), *The March of Unreason, Science, Democracy an dthe New Fundamentalism*. Oxford: Oxford University Press.

Tester, K. (1991), *Animals and Society: The Humanity of Animal Rights*. London: Routledge.

Thirtle, C. (2003) Can GM technologies help the poor? The impact of Bt cotton in Makhathini Flats, KwaZulu-Natal. *World Development* 31 (4) 1959–1975.

Thomke, S. and Von Hippel, E. (2002), Customers as Innovators: A New Way to Create Value, *Harvard Business Review*, April, 80 (4): 74–81.

Todorov, T. (1966), Les catégories du récit littéraire, Communications n°8, Paris, 125–151.

Toffler, A. (1980), *The Third Wave*. London: William Collins and Son.

Tollman, P., Guy, P., Altshuler, J., Flanagan, A. and Steiner, M. (2001), *A Revolution in R&D: How Genomics and Genetics are Transforming the Biopharmaceutical Industry*. Boston: The Boston Consulting Group. November.

Torremans, P. and Holyoak, J. (1998), *IP Law*, 2nd edition. London: Butterworths.

Triggle, D.J. (2003), Medicines in the 21st Century, or Pills, Politics, Potions, and Profits: Where Is Public Policy? *Drug Development Research*, 59: 269–291.

Turner, B. (1984), *The Body and Society*. London: Sage

UNAIDS and WHO (2004), *'3 by 5' Progress Report*. Geneva: UNAIDS and WHO.

UNAIDS and WHO (2005), *Progress on Global Access to Antiretroviral Therapy: An Update on '3 by 5'*. Geneva: UNAIDS and WHO.

UNAIDS/WHO (2004), *AIDS Epidemic Update 2004*. Geneva: Joint United Nations Programme on HIV/AIDS (UNAIDS) and World Health Organisation (WHO).

Unger, J.M. (1987), *The Fifth Generation Fallacy*. New York and Oxford: Oxford University Press.

Vaidhyanathan, S. (2001), *Copyrights and Copywrongs: The Rise of IP and How it Threatens Creativity*. New York and London: New York University Press.

Van Dijk, J. (1999), *The Network Society*. London: Sage.

Varela, F., Thompson, E and Rosch, E. (1991), *The Embodied Mind: Cognitive Science and Human Experience*. Cambridge, MA: MIT Press.

Vass, S. (2004), TV Calls on Interactivity, *Sunday Herald*, 29 August 2004. Available at http://www.sunday herald.com/print44339 (accessed 25 October 2004).

Wagner, E.L., Scott, S.V. and Galliers, R.D. (2006), The Creation of 'Best Practice' Software: Myth, Reality and Ethics, *Information and Organization*, 16: 251–275.

Waldby, C. (1996), *AIDS and the Body Politic: Biomedicine and Sexual Difference*. London: Routledge.

Walker, K. (2000), Public Service Broadcasting and New Distribution Technologies: Issues of Equality, Access and Choice in the Transactional Television Environment, in S. Wyatt, F. Henwood, N. Miller and P. Senker (Eds.) *Technology and In/equality: Questioning the Information Society*. London: Routledge.

Ward, K., Davis, M. and Flowers, P. (2006), Patient 'Expertise' and Innovative Health Technologies, in A. Webster (Ed.) *New Technologies in Health Care: Challenge, Change and Innovation*. London: Palgrave Macmillan.

Watney, S. (2000), *Imagine Hope: AIDS and Gay Identity:* London: Routledge.

Webster, A. (2002), Editorial: Risk and Innovative Health Technologies: Calculation, Interpretation and Regulation, *Health, Risk and Society*, 4 (3): 221–226.

Webster, F. (1995), *Theories of the Information Society*. London and New York: Routledge.

Weick, K. and Browning, L. (1986), Argument and Narration in Organizational Communication, *Yearly Review of Management of the Journal of Management*, J.G. Hunt and J.D. Blair (Eds), 12 (2): 243–259.

Whatmore, S. (1999), Hybrid Geographies: Rethinking the 'Human' in Human Geography, in D. Massey, J. Allen and P. Sarre (Eds.) *Human Geography Today*. Cambridge: Polity.

Whinston, A.B., Stahl, D.O. and Choi, S-Y. (1997), *The Economic of Electronic Commerce*, Indianapolis: Macmillan Technical Publishing.

Wiesniewski, D. (1996), *The Golem*. Clarion Books.

Wilkinson, R. (1997), *Unhealthy Societies: The Afflictions of Inequality*. London: Routledge.

Williams, C. (2005), Raising Interactive Television Standards, *Cable & Satellite International*, July–August 2005, Available at http://informitv.com/opinion/2005/07/raisinginteractivetelevision/

Williams, M. (2003), Target Validation, *Current Opinion in Pharmacology*, 3: 571–577.

Woodward, K. (Ed.) (1980), *The Myths of Information: Technology and Postindustrial Culture*. London: Routledge & Kegan Paul.

Wyatt, S., Henwood, F., Miller, N., and Senker, P. (Eds.) (2000), *Technology and In/equality: Questioning the Information Society*. London: Routledge.

Wynne , B. (1996), May the Sheep Safely Graze? A Reflexive View of the Expert-Lay Knowledge Divide, in S. Lash, B. Szerszynski and B. Wynne (Eds), *Risk, Environment and Modernity: Towards a New Ecology*. London: Sage.

Yourdon, E.N. (Ed.) (1979), *Classics in Software Engineering*. New York: Yourdon Press.

Zuboff, S. (1988), The Age of the Smart Machine: The Future of Work and Power. New York: Basic Books.

Index

General Editor: *Steve Jones*

Digital Formations is an essential source for critical, high-quality books on digital technologies and modern life. Volumes in the series break new ground by emphasizing multiple methodological and theoretical approaches to deeply probe the formation and reformation of lived experience as it is refracted through digital interaction. **Digital Formations** pushes forward our understanding of the intersections—and corresponding implications—between the digital technologies and everyday life. The series emphasizes critical studies in the context of emergent and existing digital technologies.

Other recent titles include:

Leslie Shade
    Gender and Community in the Social
    Construction of the Internet

John T. Waisanen
    Thinking Geometrically

Mia Consalvo & Susanna Paasonen
    Women and Everyday Uses of the Internet

Dennis Waskul
    Self-Games and Body-Play

David Myers
    The Nature of Computer Games

Robert Hassan
    The Chronoscopic Society

M. Johns, S. Chen, & G. Hall
    Online Social Research

C. Kaha Waite
    Mediation and the Communication
    Matrix

Jenny Sunden
    Material Virtualities

Helen Nissenbaum & Monroe Price
    Academy and the Internet

To order other books in this series please contact our Customer Service Department:

    (800) 770-LANG (within the US)
    (212) 647-7706 (outside the US)
    (212) 647-7707 FAX

To find out more about the series or browse a full list of titles, please visit our website:

    WWW.PETERLANG.COM